Birdland

Taplinger Publishing Company·New York

Birdland

The Story of a World Famous Bird Sanctuary

LEN HILL and Emma Wood

Line drawings by Paul Nicholas

To Mummy, Rosemary, Richard and their families.
Also Margaret who is no longer with us, but in memory
stronger than ever.

First published in the United States in 1976 by
TAPLINGER PUBLISHING CO., INC.
New York, New York

Library of Congress Catalog Card Number: 75-37334
ISBN 0-8008-0747-2

Contents

List of Illustrations

Colour photographs

Jacket illustration by Eric Hosking

Black and white photographs

Line drawings

(Except where otherwise stated all line drawings are by Paul Nicholas)

Foreword by Johnny Morris

As a very young feller-me-lad I spent hours at home with my family discussing such grave topics as the possibility of life after death and I personally always plumped for the idea of reincarnation. To return to this earth as a lion, a golden eagle or a Derby winner seemed a most appealing prospect. I don't think I ever considered being a tarantula or a giant clam or a spitting cobra! Of course, scientists make mock of such speculation, but I must confess that the theory has remained with me throughout life. Have you never met someone whom you feel sure must have been on this earth before? Have you never known a particularly clever animal which couldn't possibly have accumulated so much know-how in its own short dog, cat or horse lifetime?

Similarly, I find it quite an amusing exercise to speculate on what forms both people and animals might have taken in a previous life. And, though this may seem a somewhat strange way to approach the story of a man who is still very much alive, I'm sure you will follow my reasoning as soon as you read this book. For there is obviously no point in pondering long on what shape Leonard Hill took when he first visited this planet hundreds of years ago. He was, of course, a bird. It is impossible to say what particular bird since he has been back so many times as so many different types. How else could he know so much about all the birds that live with him? How else could he know about the flowers and shrubs that are such an essential part of the birds' well-being? Of course he's been here before.

There are many people who look after animals successfully and are very jealous of their secrets. They won't let on. In this book Leonard too does not divulge many secrets, simply because he has none. It's true that he'll tell you how to make a light lunch for a hummingbird, because that is rather an exceptional meal to make, but things which to us may seem deep and deedy secrets are, to Len, instinctive behaviour.

I think that the full force of this is illustrated when he says that no matter how many birds you have in your care you must never ignore one of them as you pass by. Every bird has his eye on you and expects acknowledgement and reassurance that you are his firm friend, that he can depend on you to the very end and that you will consider his needs before your own. As Len says time and again, you must have their complete confidence. It is a simple enough thing to say but an incredibly difficult state to achieve when you consider the delicate and sensitive personalities of some birds, their natural fear of man and their frantic reaction to sudden movement and harsh noise. 'Just gain their confidence,' says Len Hill, 'and you are on the road to success.'

Just gain their confidence! To watch Len handle his African grey parrot Juno is a perfect illustration of what he means. She lies in the palm of his hand, on her back, her legs in the air, with pale straw-coloured eyes watching the world. She knows that she is going to be tossed in a blanket. Up she goes, and down on her back, legs sticking up, perfectly contented. Up and down, up and down. It's like playing with a contented child. Just a trick perhaps. But then, how do you achieve it?

I've walked around quite a few zoos in my time and I know only too well the awful 'mopes' that afflict so many animals. They are adequately fed, watered and cleaned out, but they lack a few important ingredients—stimulation, interest and affection. To walk round Birdland on your own is to sense that you have a lively audience, interested in themselves and interested in you. To walk around Birdland with Leonard Hill is like walking into a great children's tea party with Father Christmas. There can only be one explanation for my money—he has been here before.

Introduction by
Sir Peter Scott, CBE, DSC

I was delighted to be asked to introduce this book because I have always been thrilled by the imaginative concept of Birdland. In my travels round the world I have seen quite a number of bird collections and zoos, and nowhere have I found a more complete mastery of the subject than there is at Birdland. It all stems, I believe, from the fact that Len Hill loves his birds and has a rare understanding and communication with them. It is reflected in the way that his remarkable bird garden is laid out and maintained. It is demonstrated when he walks among the birds—he knows them all by name and they know and trust him.

But Len Hill does not rest on the laurels of his magnificent creation at Bourton-on-the-Water which has given pleasure to hundreds of thousands of people. In the very early days of the latter-day Noah's Ark, the World Wildlife Fund, Len was one of the people who understood very quickly the essential point that if we care about the world's wildlife and the wilderness in which it lives, and care that it should continue to be there for the enjoyment of future generations of mankind, we have a responsibility to do something about it. He started off by collecting the pennies which were thrown into his penguin pool and making them available to the World Wildlife Fund. This brought home to us the tremendous potential of wishing wells in aid of our charity and became the prototype for World Wildlife wishing wells in zoos, bird sanctuaries and animal collections throughout the length and breadth of the country.

Len made good use of the captive audience of young people visiting Birdland, to educate the next generation and inform them of the problems which face wildlife conservationists. He built a lecture room to add to all the other splendid facilities at Birdland, in which he shows films taken by himself and other wildlife photographers. It is a demanding operation in that today's youngsters are well informed and becoming deeply interested in all aspects of animal life.

Len Hill's own horizons are global. He knows well that the birds which he loves and which give him and others so much pleasure must have their habitat safeguarded in perpetuity if they are to continue to have a place in the sun. For that purpose, when the chance came along, he bought the Jason Islands in the Falkland Island Group—two uninhabited islands where conscientious wardening ensures the peace and solitude needed by the large populations of penguins, albatrosses, steamer ducks, Magellan geese, kelp geese, elephant seals and southern sea lions.

I am happy to have this chance to salute my friend Len Hill—a gentle and warm-hearted man who cares about the world's wild creatures and wildernesses—and shares my optimism that man will solve the grave problems which beset him at this time. We have to be sure that when he does there is still a world of nature in existence to enrich the quality of his life. We, of this generation, have a responsibility to keep the options open for the next.

Good luck to this book and to Len Hill's far-sighted philanthropy.

We Made a Garden

Often, on a dark winter's morning, I draw back the bedroom curtains on a scene which would make most people rub their eyes and wonder if they were dreaming: below, on the snow-covered lawn, a silent group of penguins gaze expectantly upwards as if to say, 'Hello, are you getting up today?'. By midday, when the ice has melted, their place will be taken by a gaggle of bright, noisy flamingoes, executing what looks like a tribal dance in slow motion. Finally, as dusk falls, I call out of the study window and in flies Juno, an African grey parrot, who nestles on my shoulder to say goodnight before settling down to sleep in the house.

Such incidents, in an English garden, in the heart of the gentle Gloucestershire countryside, are part of my daily life. Yet strangers, visiting our Cotswold show-place village of Bourton-on-the-Water for the first time, are amazed to see the gaudy purple, blue, green and orange plumage of macaws flying free over mellow stone cottages and the surrounding green fields. Most people confess to being so intrigued that they make a point of finding out more about Birdland, my zoological garden housing some 1,200 birds from about twenty countries of origin, happily co-existing in the five-acre grounds of the Tudor farmhouse which is my home (see page 17).

So established is the garden now, both as a place of pilgrimage for amateur ornithologists from all over the world and as a day's outing for families, schoolchildren and old people in this country, that it is sometimes difficult for me to remember that Birdland has only been in existence less than twenty years, the realisation of a boyhood dream over sixty years ago to create just such a sanctuary in my native surroundings. As a child, school meant little to me: I was interested only in nature and what I could learn exploring our local Windrush valley. Although my parents were poor they encouraged this leaning, buying me books on horticulture and aviculture for which I paid back a little whenever I could afford it. Why, they even allowed me to do 'research' on the kitchen table, cutting up dead birds and skinning them to discover the cause of death. And I always had a tame collection of snakes and small mammals such as stoats and weasels, all of which taught me vital lessons in living with animals.

My favourite companion at that time, and for many years to

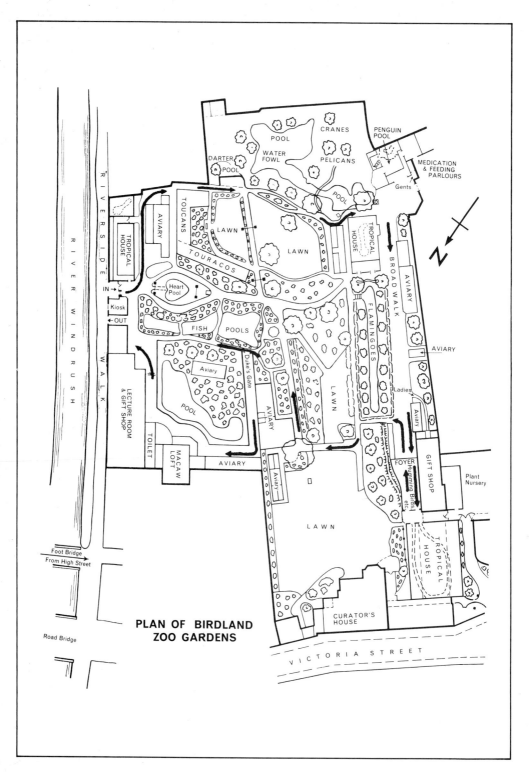

POOL

CRANES

PENGUIN POOL

WATER FOWL

PELICANS

DARTER POOL

MEDICATION & FEEDING PARLOURS

Gents

POOL

TROPICAL HOUSE

R I V E R S I D E   W A L K

AVIARY

TOUCANS

LAWN

TROPICAL HOUSE

BROAD WALK

AVIARY

TOURACOS

LAWN

IN →

Heart Pool

Kiosk

← OUT

FLAMINGOES

AVIARY

FISH POOLS

Duke's Gate

Ladies

AVIARY

Aviary

LAWN

R I V E R   W I N D R U S H

POOL

AVIARY

Aviary

FOYER

GIFT SHOP

Plant Nursery

LECTURE ROOM & GIFT SHOP

TOILET

MACAW LOFT

AVIARY

Aviary

Humming Birds etc.

LAWN

TROPICAL HOUSE

Foot Bridge From High Street

PLAN OF BIRDLAND ZOO GARDENS

Road Bridge

CURATOR'S HOUSE

V I C T O R I A   S T R E E T

come, was Joey, a humble pigeon which I raised virtually from birth. In those early days my father, a groundsman, had been employed on a large estate where the owners took a great interest in his small son who seemed to live only for birds and flowers. Indeed, my earliest recollection of learning about wildlife is on this same estate, aged four, dressed in a velvet suit with a large white collar and holding a little cane which I used to point out and name the various herbacious border plants to guests. A few years later, when my family had moved to Bourton, I cycled over to see my kind friends at the big house and was presented with a week-old squab pigeon which had fallen out of the nest and was in danger of not surviving.

The little scrap of barely pulsating life meant more to me than if I had been given a gold sovereign. I asked for a used flour bag, turned it inside out, carefully placed the pigeon inside and cycled slowly home with my precious charge suspended from the handlebars. I named her Joey unaware at the time that she was a hen bird. Rearing her in those first weeks wasn't easy and I was soon putting into practice all my seven-year-old ornithological theories. I'd read, for instance, that an ancient method of feeding pigeons was by first chewing corn to soften it so that the baby birds could then take it by mouth from their human 'parents'. The Roman legions apparently used this method to provide a constant source of fresh meat for the troops on long marches. All the prisoners captured en route were employed as corn chewers to feed and fatten up the squab pigeons which were thus able to be taken early from the nests, so that the adult birds laid again and the process was repeated. I fed Joey in the same way and, in a few days, she learned to take the corn from my mouth, which made her very attached to me as I had taken the place of her mother.

Rewarding as this devotion was for a small boy, making me the envy of all my friends, it also got me in to some serious scrapes. Like many young lads in rural communities I sang in the choir and was always in attendance for important events such as the christening, one hot summer's afternoon, of a local bigwig's baby. The heat became so unbearable during the service that the vicar opened a side door in the chancel and, as we were all singing solemnly away, in breezed an uninvited guest—Joey. On to the reredos she flew, up and down the aisle, before finally spotting me and coming to rest in the pew. All the choristers were tittering in amusement and I must confess that a little light relief did us all no harm, although the congregation were not so pleased and the canon administered a sharp reprimand when he caught me alone. Strange how times change, isn't it? I always think that today someone would be only too pleased to deliver a sermon on such a story. Joey, I'm

delighted to say, never realised the nature of her crime and lived happily with me for another eighteen years.

By what seems a fated coincidence, we were living then in a small cottage attached to the stable yard of Chardwar, the old Elizabethan manor which is now my home. Water was scarce and we were limited to two bucketsful a day which I had to collect from the pump outside the door of the main house. How little I thought then, as I was frequently shooed away by the servants for making too much noise with the pump handle, that some fifty years later I would be living in that same house surrounded by my beautiful birds. Yet oddly enough George F. Moore, the famous orchid grower and local philanthropist who owned Chardwar at that time, contributed in some small way to my present success by helping foster what I suppose can only be called my business instincts. Often pausing for a chat as he came across me caring for my bird companions, G.F.M. played a special game with my tame jackdaw. He would secretly roll a sixpenny piece on the floor and cover it with the end of his walking stick, then dare the bird to find it. Of course, the jackdaw always succeeded and, watching where he alighted with his booty, I was thus able to increase my pocket money by some two hundred per cent!

All through my boyhood my only interest was wildlife. I was never very clever at school as there always seemed to be so many more interesting things going on outside the classroom, and the only lessons which really held me were geography and drawing. I was always pleased when the bell rang at the end of

(Opposite) Getting to know a newly arrived rockhopper penguin

Chardwar manor today with flamingoes on the lawn in the foreground

17

the day, allowing me to get out in the fields where I felt most at ease. As I grew older I became more interested in conservation, although I doubt if I'd even heard the word at that time, let alone knew the meaning. Like many country boys I was very proud of my collection of birds' eggs, but I also became a friend of all the local gamekeepers, keeping them informed of birds which I discovered shot or poisoned in the local fields and woods. This was long before the days when protective acts were introduced to guard certain species, and a peregrine falcon which I found killed is still to be seen, stuffed, in a local museum.

Sadly, these happy early teenage days and interests all too soon had to take second place to work as I left school at fifteen to be apprenticed as a carpenter. My father paid £25 for me to have the 'privilege' of earning the princely sum of 5s a week over a five-year period: and I had to work fifty-six hours a week, cycling twelve miles a day, in order to obtain it. Although I grumbled at this new brake on my freedom, looking back I am thankful for being put to work in a strict school where the old virtues of diligence and discipline helped to mould my character. I shall never forget my 'guvnor' sacking a reliable employee of some twenty years just because the man had the audacity to smoke during working hours! Yet, if he was tough, the old man was also fair, with his own peculiar sense of humour. My very first task as a journeyman, rehanging a door, was disapproved of by the lady of the house. At my second attempt, the door was perfectly hung, but still she complained to my boss that the wood had a lot of knots in it. 'God bless my soul, woman,' he replied, 'do you buy your meat without bones in it?'

At the beginning of my apprenticeship the company had been a prosperous one employing over eighty but, by the time I was ready to receive my first week's wages, we were in the middle of the 'thirties depression and only a dozen men remained. Being the junior member of the crew I was ever fearful that I too would soon be joining the dole queues and, when the request came for me to collect my money first, at twelve o'clock on the dot, I was sure that I was about to be sacked. Feeling destitute, I was amazed to hear my employer say: 'Laddie, I've got a fortune for you, more than I've ever paid a boy before. Here's 1s 2½d an hour—and £5 from Henry Ford for the work you've done for him.' I'd never had so much money in my life and felt as if I could have bought the street! Ford, the great American industrialist, had purchased two old Cotswold cottages of the type in which his English ancestors would have lived. First we restored them, then dismantled them, stone by stone and slate by slate, all carefully numbered for shipment to the United States. The dwellings were then re-erected, along with a blacksmith's shop, at the village of Dearborn, near Detroit. Today, having travelled the length and breadth of North America, showing films of my travels and adventures to aviculturalist societies in such cities as New York and Chicago, and making many transatlantic friends, I frequently remember their fellow philanthropic countryman.

All this talk of my early life may seem to be digressing a little from the story of how Birdland came about but, in fact, it is really an integral part of that story, as I would never have been able to contemplate the venture were it not for a lot of hard work as a young man and several lucky breaks throughout my career. I suppose that one of the turning points of my whole life was when, at the age of eighteen, I had two fingers chopped off in a planing machine, and sued the company for what would now be termed industrial damages. The day of the court case was terribly foggy and the defendant's solicitor failed to appear in time for the hearing. Indicating that a sum of £75 seemed adequate compensation, the judge then adjourned the case until the arrival of the insurance company's solicitor. When he eventually did turn up, over an hour late, the case reopened with the judge in a very bad mood. Admonishing the counsel for being late when he, personally, had managed to get to the court on time, despite the bad weather, the judge quickly suggested that I should be awarded £100 damages—the quickest £25 extra I have ever earned through doing nothing! As the money had to be invested on my behalf I suggested adding a small sum to it in order to buy a hundred pounds' worth of war bonds, a plan which seemed to tickle the judge greatly.

I mention this as a turning point in my life as the bonds, together with the additional interest, enabled me to get married and buy a house costing £420, for the sum of £42 down and a twenty-one year repayment mortgage of 12s a week. I well remember giving Nell, my wife, my weekly wage of about 45s from which she used to pay back the building society, buy all our food and clothes and be delighted if she had 1s 11d over to indulge herself with a pair of silk stockings. Eventually, however, I too became redundant and we had to exist on only 16s a week dole money, making do with whatever jobs were going, such as haymaking and labouring. This seemed a pretty poor state of affairs to me and I decided that I could surely earn more than this working for myself doing odd jobs around Bourton. The clerk at the labour exchange was so amazed at this initiative that he shook hands and wished me luck, declaring that I was the first person ever to have proposed such an apparently revolutionary scheme.

Naturally, being self-employed was an uphill task at first, but

I circularised every house in the district that I was available for all kinds of jobs at 1s an hour and, amazingly, I very soon had more work than I could cope with, so that by the end of the first year's business I was employing two other men. We were then building houses for a local landowner who, in the middle of a conversation one day, mentioned that he was selling a parcel of land to one of the largest construction companies in the country. Angered by his thoughtlessness in apparently not caring that large numbers of jerry-built little boxes would spoil the entrance to our beautiful village I spoke out so strongly against the idea that, the next day, he asked if I would like to purchase the land to erect buildings which would blend in with the area. Of course, I had no money to effect such a deal, but we eventually reached a compromise whereby I paid £100 for the option to build for three years. As each plot was sold, I would pay back £77 and, at the end of three years, my original £100 deposit would be returned. It seemed an excellent scheme but I thought it best to let my solicitor see the contract as I had never done anything so ambitious before. This was in 1939 and the international situation was troubled, so my solicitor suggested that we inserted a clause to the effect that, if hostilities commenced, I would be allowed the same unexpired portion of the lease after the Armistice. What wonderful advice that proved to be; by September we were at war and only the very first house was nearing completion when all private house building was stopped by law.

Throughout this period I had never ceased to be interested in birds and the idea of forming a small, private collection of unusual species was always at the back of my mind. The growing responsibilities of family life, with the birth of three children, forced me to concentrate on the more immediate problem of economic survival, especially now that I had a plot of land on which I could not build. How best to utilise it was the problem. As Bourton is only a few miles away from the then highly important air base at Little Rissington, it occurred to me that I could not go wrong in providing food for the troops stationed there, and so I planted 12,000 cabbages. When these were ready I asked in the village if anyone was interested in taking this amount of cabbages, only to be offered the miserly sum of £5 by a local greengrocer. He was apparently serious and even put the offer in writing, a letter which I have treasured to this day. As I'd hoped, however, the NAAFI bought all the cabbages and asked me to supply them with lettuce, offering to pay at the rate of 2s a dozen or 6d a pound. I accepted payment on the basis of weight, growing Webb's White Wonder, a beautiful, big lettuce which thrived in our lovely soil and, of course, weighed even heavier when cut in the early morning

An East African crowned crane, often seen on the lawn with the flamingoes

with the dew still on it so that I could deliver to the camp by 5am!

This was a time for taking advantage of every opportunity to keep a decent standard of living and I would come back from the base with all the kitchen waste material to feed to my pigs as we were allowed to kill two a year under licence throughout the war years. Several of my building labourers continued to be employed as we had obtained a government contract for producing wooden bomb-nose caps for use in the Far East, and we adapted our woodworking machinery so that we could produce thousands in no time at all. The main contractors were so intrigued with our output that they came to photograph our methods and called in the War Department. Unfortunately we were apparently producing large numbers in half the time taken by other people so our rate for the job was also halved! After this we supplied over a million wooden coat hangers for US troops stationed in this country and continued with lots of other small jobs, still growing and selling vegetables, and we were so successful in this field that I even expanded later into a small nursery business.

At the end of the war building was still restricted and I was always short of money, but slowly began to prosper. I have always been fascinated by large old houses, particularly because of their historic interest and because I could imagine how they would look when converted for present-day living. At this time we were working on estates owned by a prominent London financier who would drive down to Gloucester after a day's work in the City to discuss work on his buildings with me. Money seemed of little consequence to him and he would frequently hand me sums of up to £1,000 from his safe to pay for particular projects. I admired him greatly and was more than pleased to take his advice when he offered to introduce me to a famous banking house which would loan me money to enable the business to expand. On his security I was loaned £5,000 on the value of my property, a sum which seemed unbelievable at that time, and which almost brought about my downfall.

As the business went from strength to strength I began to build up more debts, especially as by now my financier friend had stopped paying me for work on his property. Eventually the day arrived when I owed £6,000 and my credit limit was £1,000 below this. After various fruitless phone calls I eventually went in person to my 'benefactor's' local home, only to be told by the butler that he had left that morning for the Continent and was not expected back for some time. For the next two months we took on every available job where we could ask for immediate payment but even so, at the end of every week, I had to ring my bank manager to explain that I did not have enough money to pay the wages. Fortunately he was understanding and never once let me down in making up the necessary sum so we managed to struggle on week by week.

Then, one wet and windy October night, after regular phone calls to my client's London home, I established that he was back in the country and would be in for dinner. Leaving no message, I drove straight up to the capital at the end of the day's work and, by 9pm, was ringing his doorbell. As was only to be expected he declined to see me but, equally true to character, I declined to be put off. Buying a pile of newspapers from a passing boy I made myself as comfortable as possible and prepared to spend the night on the doorstep. The local policeman was prepared to move me on for loitering until he heard my story, then he wished me luck and went on his way. After a night of little sleep I was more than pleased to accept the butler's early morning invitation to go inside, below stairs of course, and warm myself with a cup of tea. As Mr Moneybag's own early morning tea was ready to be delivered to him I wrote a little note to send up with it: 'Dear Sir, I am waiting here until I have my money. This is your duty and my just reward for my labours.' Within minutes back came the reply that I must remove myself from his premises forthwith. At that moment in came a footman with a pair of shoes which he had just finished cleaning. He went into the lift and I followed. Up we went and, without speaking, got out. Along the corridor we walked in stately file, looking rather like my penguins do today, and in we both walked to the master bedroom. Of course there was a great argument, although I refused to lose my temper and it became obvious that I was not going to get anything out of him. However, I took the opportunity to memorise from a paper lying on a chest the name and address of a company of architects which he used and resolved to see if they could help me.

By the time I found their Sloane Street offices I had seen more of London than ever before. It seemed at first that my visit there would also be in vain as our financier friend was heavily in debt to this company too. However, when I explained that I would be bankrupt if no action were taken soon, the senior partner offered to act for me in the dispute and was also kind enough to come down and look at my building, drawing up a most favourable report for the bank on the quality of my work. This gave them the necessary assurance to continue with the loan until our debtor could be taken to court and, after the case went against him, payments began to filter slowly through, though I still had no ready capital.

Later, however, the tables were turned with a vengeance, when one of his estates, for which he had not paid a penny, was put up for sale. All was sold apart from the huge main house

General view of the garden in the early days

which was withdrawn at the then seemingly very high price of £1,800. With no money, but visions of what could be done with the building, I telephoned the bank manager and filled his ear with details of how I could convert the house into two properties as well as making four cottages out of the stables. On the promise that I would complete the work within twelve months he offered to put up the capital for the project and, by the next morning, the house was mine. We worked night and day on that estate, never sparing ourselves, and using only the very best materials. It was a gamble which certainly paid off. By the end of that year I had sold the houses for nearly £18,000, paid off my bank loan, gained confidence in my business ability and, more important, was beginning to be in the position where I could look around for a place with enough land to allow me to create my dream of exotic birds living, breeding and flying free.

By the mid 1950s I knew that the time had come to make my dream a reality. By happy chance, Chardwar, the beautiful old house which I knew from my boyhood, was for sale and I determined to raise the money for its purchase. On a never-to-

be-forgotten day in 1956 the house became mine and we set about clearing the garden before moving in. What a task that was! The previous owners had found labour difficult during the war years so had simply dumped all their rubbish round the edges of the grounds. During those first few months we took 150 tons of ashes and cans out of the garden before we were able to begin landscaping, erecting aviaries, converting the old orchid greenhouses and digging pools.

Initially I planned the garden for my private pleasure with the idea of retiring early and devoting myself to my hobby. As the place took shape my concept of its function changed and I thought that it would be a good idea to open the garden to the public, not specifically as a money-making venture but simply to help with the upkeep, as well as bringing a little of my own enjoyment to what I then thought of as a handful of visitors.

Amusingly, my plan created a certain amount of opposition in our somewhat conservative village. During reconstruction the garden gates were boarded up and, as the diggers could be seen at work excavating, a rumour grew up that we were building a swimming pool. This apparent scheme so angered

The ornamental pool and summer house showing the importance of the correct garden environment with suitable plants and shrubs

some local worthies that a petition was raised to stop the building of this supposed pool. As I too did not want a public swimming bath in my garden I whole-heartedly supported this move and even obtained a copy of the petition so that I could add my name to it!

As the day grew nearer for the garden to be on view I decided that it would be a nice gesture to invite Bourton's old-age pensioners to become honorary members so that they could be admitted free whenever they wished to meet together and wander round or sit in the peace of the garden. I then approached a local GP who had supported the petition against the pool and described my plan to him as he was on the committee of the village Cheery Club, the old people's social society. Imagine his embarrassment at learning that I had made a *bird* garden! But he was pleased to accept my suggestion on behalf of the pensioners. I too was pleased because, from the day Birdland opened in June 1957, I had seventy extra pairs of eyes helping to police the grounds. It still gives me special pleasure to see old people from much farther afield enjoying the garden today. They gain admission at reduced price, like children, and countless numbers return each year to see the birds and flowers.

In our first year of operation, on just over three acres with only thirty different species and about two hundred birds in all, to my astonishment the garden attracted over 30,000 people. Today, with more land, more sophisticated methods and rarer breeds, we are a commercial enterprise with over 700,000 visitors annually. I no longer want more birds, simply to rear varieties in danger and thus preserve them for succeeding generations. I have been offered millions of pounds to sell out, to set up bird zoos in different parts of the world, but my work is here, in this garden, the small boy's dream-come-true, which gained its very name from one of my childhood book purchases, Oliver G. Pike's famous ornithological work, *Birdland.*

An Eye for a Bird

When selecting the garden's inhabitants I determined to do
things correctly and achieve a balance between the birds and
their environment. I suppose that I could be classed as a poor-
man's plant collector too, as I always bring back flowers not
found in these islands from my foreign trips, in the hope that we
can raise some uncommon species in our temperate climate.
From the beginning then, I took great care in working out
which birds would live happily with which plants, so that each
would benefit from the other. Too many gardeners just grow as
many different plants as possible, and I'm afraid that the same
principle is true of many zoos. Overcrowding makes it almost
impossible to keep such places disease-free and, as anyone with a
pet will appreciate, cleanliness must be one of the most
important considerations in running an organisation like
Birdland. Each inmate has a sterilised feeding bowl put out
daily, a vital measure, as any epidemic could be fatal with a
stock concentration such as ours.

Many hours are spent on the correct preparation of food, a
formidable task when you consider that half a hundredweight of
fish is put out every morning for the penguins alone. We order
some seven tons of herrings and shrimps from the Lincolnshire
docks, each year, keeping about half a ton always in stock. This
is preserved in a deep freeze, then transferred to a special
refrigerator, set just above freezing point to allow the fish to
thaw out slowly, an essential precaution to prevent its becoming
soapy and unfit for consumption. Sprats too are bought in six-
ton lots which last for approximately six months. These come
down from Aberdeen in cold storage tankers and undergo the
same thawing-out process. We feed some whiting to the birds
but it is usually only possible to obtain this fish gutted and we
prefer ungutted varieties which enable the new birds, in
particular, to establish a good digestive system early on. Live
trout from local hatcheries are especially good for this purpose
and also for tempting ailing birds which may be off their food
for a time.

It is quite a business too just making sure that there is a
constant order placed with all our sources round the world and
checking up on the way in which stocks are being consumed.
We feed gallons of honey, mainly imported from Rumania,
Australia and New Zealand, to such birds as the lorikeets. The

A member of staff going about
the daily business of preparing
food for distribution

honey is usually obtained in ton lots from wholesalers in London and brought here on long-distance parcel service, although we often make trips to Heathrow airport especially to pick up food which we've ordered direct from foreign suppliers. Honey is usually fed to the birds in a melted form, mixed with two tablespoons of malt in a quart of warm water, making a special kind of artificial nectar. Sometimes we mix it with spoonsful of Complan, a vitamin-packed food based on dried, skimmed milk, and a few half-inch cubes of brown bread. The lorikeets really thrive on this, so it's a good recommendation for the branded food for humans too, I suppose!

Fruit is the birds' other main food. Cases of apples, bananas and grapes as well as nuts and sunflower seeds by the hundredweight are consumed weekly. It is vital to maintain a good relationship with our suppliers as the fruit is flown in from all over the world—Cyprus, Spain, Italy and South Africa—depending on when it is in season. And where else would the nuts come from but Brazil? Both nuts and fruit are washed, then cut up into suitable bite-size pieces before being distributed.

It is surprising to many people that meat does not play a

A great Indian hornbill

Easily recognised by its greatly swollen, elongated bill, the toucan comes from South America. The immense bill, of ornamental rather than practical value, is filled with air, the thin outer shell being covered in a tough skin

greater part in many birds' diets, but if they stopped to think they would realise that it is mainly the birds of prey which live mostly on meat in their wild state. In all we use up to about twenty pounds of meat weekly, mainly bullocks' hearts imported from New Zealand at exactly half the price it costs to buy them in this country, even including the carriage from half way across the world and the dealer's percentage! Toucans and hornbills are both meat-eaters and we mince portions of meat up finely to feed to the smaller birds such as owls and pittas.

Some live food is essential in the diets of several varieties of birds. The owls, for instance, unable in captivity to catch the majority of their own food as they would if entirely free, consume a hundred-weight (112lb) of day-old chicks a month which we buy from a local hatchery. Hornbills too include about half a dozen live mice or, if these are not available, a pound of minced beef, along with their daily ration of eight pounds of fruit. We also buy maggots by the gallon and about thirty pounds' worth of locusts are purchased every week from a Cheshire breeder to be fed to the delicate tropical birds such as the quetzal.

On one occasion a disaster had occurred in the breeding of mealworms, a hundred pounds' worth of which are flown in

monthly for us from Germany. This nearly meant a disaster for Birdland too as no mealworms could be found at short notice in this country and I was desperate for food for the bee-eaters. We searched all sources of supply drawing blanks everywhere so that, finally, I rang the BBC and told them of my plight. They kindly allowed me to broadcast an appeal for food which produced a fantastic response; within twenty-four hours we had information enabling us to collect about 10,000 bees and 17 wasps' nests. And each year, in August and September, we trap about 200 live wasps and bees a day which the bee-eaters catch on the wing and swallow whole after knocking them senseless, just like kingfishers with their insect prey.

Preparing all this food is very time-consuming and we now have three people helping in the kitchen, not to mention offers of unconstructive aid from the macaws who create chaos when on the look-out for a sample of all the other birds' food! In the early days, however, we only employed a woman part-time to help with the washing up, the majority of the other work being done by my younger daughter, Rosemary, who gave up her job at the Foreign Office to be at Birdland full-time. Those with long memories of the original version of the TV panel game 'What's My Line?' may recall the astonishment of Cyril

A spotted wood owl from New Guinea, one of the Birdland inhabitants which requires live food in its diet

A flamingo neatly trussed up in a stocking, ready to make a safe journey

(*Above right*) I often think that these two palm cockatoos look like mourners at a funeral, blushing at being caught out telling a joke

(*Below right*) The odd-looking Papuan hornbill with its massive casque which appears to be more for decorative effect than any utilitarian purpose

Fletcher when teenager Rosemary beat the panel, as an aviculturalist.

Now married with children of her own and, for the first time, living away from Bourton, Rosemary was indispensable at the beginning, often accompanying me on trips to bring back new birds. Starting in a small way we purchased from suppliers in this country, then went over to the Netherlands where the dealers have a much larger variety of stock. Not many birds are subject to quarantine regulations when bringing them into Great Britain, so long as the transaction is carried out officially through the Ministry of Agriculture. We had many amusing experiences when returning home with our live companions, especially the larger birds such as flamingoes.

To transport flamingoes we 'dress' them in ladies' stockings. The toe is cut off, the stocking rolled up as if to go on a leg, then placed over the bird's head and unrolled down its back with the long legs trussed up inside. The stocking is then pulled tight over the flamingo's bottom and tied in a knot, leaving the head sticking out, which keeps the birds nice and snug whilst allowing some movement. I remember driving off the Channel boat at six one morning, with the birds on the back seat waving their heads around like inquisitive old women. I don't think that the young customs officer on duty could believe what he saw. His expression seemed to say: 'I've heard of pink elephants, but . . . !'

The elegant birds soon recover from their undignified travelling garb and prove one of the garden's major attractions

with their exciting colouring and communal, rushing movements. We have thirty flamingoes (*Phoenicopteridae* family) at Birdland at present, representing all six species. Two of these, the greater and lesser flamingoes, are known as African varieties as they hail primarily from that part of the world, although the greater flamingo is very widespread, being found in southern Europe and the Middle East as well as in parts of South America. The harsh mountain regions of the Andes are the home of the Chilean flamingo, and here too can be found the Andean and James's flamingoes, the latter being rediscovered merely twenty years ago and having an estimated world population of only 15,000. The striking roseate or Caribbean flamingo, deservedly attracts attention beside the paler, almost white European flamingo and, when seen together in motion, it seems to me that the two species resemble a constantly changing sunset.

Flamingoes are extremely ancient birds, fossilized remains dating back 40 million years having been found. Most people are amazed to learn that such exotic creatures were once natives of the British Isles. They are large, up to six feet in height, having longer necks and legs in proportion to the rest of their bodies than any other birds. Visitors to Birdland frequently observe a flamingo asleep, resting on one leg, and ask why this is so. I can't resist replying jokingly that if they don't rest on one leg they will fall down, but unfortunately some people seem to believe me! It *is* easy, however, to be misled into thinking that the flamingo must be a member of the heron family because of this trait. Odd as it may seem, as proved by anatomical evidence, the flamingo is a distant relative of ducks, geese and swans. There are similarities in behaviour, if not in looks, as flamingoes gabble together and honk when in flight. Often in the night, hearing them through veils of sleep, I imagine that a flock of geese are on the lawn below. Flamingoes are also very gregarious, breed in colonies and the young take to water as soon as they are hatched.

Yet the flamingo differs from related species, and from all other birds, in one important respect: its odd-shaped bill with the curved upper part, making it look like an imperious Roman emperor, is an ingenious food selector which man can only pause and marvel at. As the flamingo stoops to feed, the bent bill at the end of that long neck acts as a scoop, a mini dredger in which it collects mud from the bottom of pools. Special filters on the beak and bristles in the mouth then take over in a sorting out process so that only palatable food is swallowed and the rest ejected.

Their gangling appearance often causes flamingoes to be dismissed as stupid birds, an opinion which I certainly cannot

(*Opposite*) The hyacinth macaws, Leah and Mac, the garden's trademark almost, here seem to have taken a liking for Gothic architecture

substantiate. As mentioned at the beginning of this book, our flamingoes usually congregate on the lawn, in front of the house. They are also fond of a little pool in a dell, lower than the rest of the garden, (see page 52) and they have their own heated shed too in case of bad weather. So regimented are these birds in their behaviour that I can always forecast the immediate weather conditions from their actions. If the night is going to be mild but windy, then the flamingoes can always be found in the dell. However, if a cold spell is about to begin, then off they progress, solemnly, in single file, to the warm shed. I've never known their instict to be wrong. In fact, I'd recommend all weather stations to keep a few flamingoes: with respect, the forecasts might be more accurate!

It's little observations of this kind which enable us to get to know the different birds, their habits and feelings. My staff treat all the birds as individuals, respecting their likes and dislikes as we would those of humans. Have you noticed that if you walk by a bird and take no notice of it then it will ignore you also, but if your eye meets that bird's eye there is an immediate response? Just try this first step in basic communication. It is from such simple contact that understanding grows and a mutual trust is formed so that, today, my parrots and macaws have come to know my footsteps before I call out to them. And woe betide me if I forget to acknowledge a bird! If, on a busy day, I'm passing through the garden and don't have time to say hello to one of the macaws, then it will 'speak' to me first, claiming my attention by uttering loud cries and generally flapping around to make its presence felt.

We give all the birds a name which is used constantly when calling and feeding so that gradually, like a dog, the bird will answer to that name, and a relationship can then be built up. This establishment of personality and trust was an essential step in our plans to let the birds fly free. From the start I determined that the birds of Birdland should be at liberty as far as possible, to live unconstrained in a habitat very different from their native one. Now the birds respond when I call, ranging free during the day and coming home to sleep at night.

Such freedom is the result of years of training and constant patience. In the early days we allowed just two birds to go free, the blue and gold macaws, Charlie and Ann (see page 136). As soon as the birds responded to their names we took a piece of wood, two feet long and some two inches in diameter, suitably gnarled for claws to grasp, and taught them to recognise, then come to this stick, of which more later. We then released these two in the garden where they were later joined by another pair of macaws, Jack and Sheba.

Naturally the birds would not remain in the garden at first

(*Opposite*) Believe it or not this is a flamingo, drawn by an infant school pupil after a visit to Birdland. I particularly like the head of the following bird!

and we experienced many sleepless nights rounding them up from miles away. There are many distractions in the area of Bourton, any one of which could startle the macaws to take flight. When undisturbed, the birds simply fly around in circles, anti-clockwise. This seems to be some unwritten law of nature, for you may notice that bathwater runs out of the plughole in exactly the same way and runner beans twine round their sticks this way too. Any sudden noise, such as a car backfiring or a gunshot, straightaway disturbs the birds into taking flight, always in the direction in which they are facing. Being close to an air base with the constant noise of low-flying aircraft, not to mention lakes at either end of the village which attracted swans in spring, we had hardly a day in those early years when we didn't have to fetch back at least one errant inhabitant.

Macaws are very sensible birds, of immense longevity, and it seems as if real common sense has been inbred in them throughout the ages. The hyacinth macaws, Leah and Mac, (see page 35) which are almost our trademark, behaved so well when first set free to join the birds already out that we had little difficulty in finding all of them at the day's end. If either one of the pair was out of the garden, then the other would spend the hours flying out to the spot and back, out and back, so that we were able to ascertain, within a certain distance, where the other birds were. Sometimes, if it was foggy, or dark, the liberty macaws might have flown beyond calling distance so that they were more difficult to locate: a macaw can pick up sounds from two miles away, quite easily, but often they flew even beyond this range.

It was on such nights that we had to employ 'operation birdlift'. Neighbours obviously thought us mad, setting out at dusk with an old estate car full of poles, food and a trap cage to

(*Opposite*) Close-up view of the head of a blue and gold macaw, showing the pronounced facial markings which look like thickly applied, vivid make-up

The 'chimney sweep's' sticks used to rescue errant liberty macaws in the early days

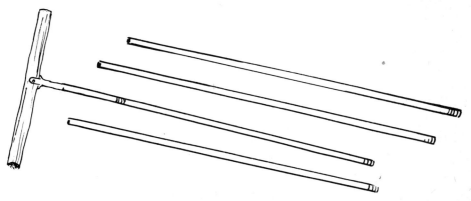

39

bring back errant birds, but most people were very helpful in ringing up to let us know that they had sighted the birds.

An essential part of our equipment was that little stick on to which the birds had been trained to fly, in the middle of which had been fixed a screw. We also carried a series of six-foot rods, rather like those used by old-fashioned chimney sweeps, which joined onto each other and finally fitted into the thread in the centre of the gnarled stick. By assembling the poles we could 'climb' fifty feet or more up trees in search of the birds. The rods would be wiggled, very carefully, through the foliage, until they came to within a yard of so of the macaw. Perched aloft, in an alien landscape, perhaps a bit insecure and certainly somewhat sleepy, the bird would suddenly see a familiar object in front of it, often holding a dainty chocolate morsel. 'Hello, there's my little stick' it would think, then hop onto the wooden 'perch', to be lowered safely and popped into the trap cage.

Not every rescue was so easy however, as often a bough would break or some other noise disturb the night air, and off the bird would fly to land miles away, possibly in an even more inaccessible place. So we began the whole process again . . . and again . . . and again . . . and many's the time I've been out all night rounding up the miscreants. Yet we have never lost a bird in the garden's history. Some have been purloined by the unscrupulous, but they are always returned safely. Luckily Birdland is so well known in the area that anyone suddenly producing an unusual bird is very quickly reported back to us and, in the same way, any exotic bird found 'out of bounds' as it were is soon back in the garden. It gives me great pleasure to feel that the establishing of mutual confidence between man and bird which has enabled these creatures to fly free is a correct step along the road towards wildlife preservation and conservation.

A Bird in the Hand

The world finally seems to have woken up to the fact that vast areas of wildlife are threatened with extinction and certain species are now legally protected. Such action may pay long-term dividends, but more immediate steps need to be taken in many cases where natural environments are already despoiled or man, the ultimate predator, has so plundered stocks for his own commercial gain that some species are in danger of disappearing altogether. I count the successful breeding of rare birds as my major positive contribution in this area, and it gives me immense pleasure to know that birds born here, in Gloucestershire, are now forming colonies in their native habitats.

One of our happiest ventures in this field has been with the tiny Hahn's macaws (see colour illustrations, page 69) from Trinidad, a species which had been extinct in the island for over fifty years, as mentioned in Herklots's authoritative book *Birds of Trinidad and Tobago*. This delightful bird (*Ara nobilis*) has predominantly grass-green plumage shading to blue on the front of the head. It has bright scarlet feathers under the wing and on the shoulder, with a dull gold-coloured tail. This rarity was, until quite recently, found only in South America and Guyana. I caught my first glimpse of it in this country as three were being carried into a pet shop and, though I did not know at the time what they were, I bought all three on the spot. At the time I considered that I paid dearly for them, but they were so unusual and have repaid our care so well that I now consider my money more than well spent. By constant surveillance during the first twelve months, with due attention paid to diet and warmth, we eventually made the birds feel so much at home that some three years later they produced their first chicks. Yearly additions have been made to the family since then, all being reared successfully, and we have now bred over thirty Hahn's macaws at Birdland with last year's youngsters still flying around at liberty here. In 1964 I was delighted to be able to deliver two pairs back to their homeland where they are now happily nesting and there is now a good chance of the survival of the species. I was interviewed on both radio and television in Trinidad which was useful in making the public aware of this successful attempt to reintroduce a native bird to their country.

I also appeared on television in Jamaica when I went to the

The materials used in the tiny intricate nests of the streamertail or doctor hummingbirds vary according to location, generally being moss and cotton

island a few years ago at the government's invitation to assess the possibilities of creating a bird garden there. Here, I fell in love with the Jamaican national bird, the streamertail hummingbird (*Trochilus polytmus*) (see page 69). To see this beautiful creature in flight is to know instantly why it gained its name, as the two long tail feathers of the adult male stream out behind it. Another popular name for it is the doctor bird, a reference to its erect black crest which resembles the top hats worn by doctors in past times, and to the smart black tail feathers which look a little like a very proper frock coat. But let me leave a fuller description of this bird to the late May Jeffrey-Smith, a dear lady and excellent ornithologist whom I had the pleasure to meet when on the island. She gave me a copy of her invaluable little book, *Birdwatching in Jamaica*, now sadly out of print, from which this extract is taken:

It is indeed a beautiful creature as it darts from flower to flower, sometimes pausing to puncture the base of an acanthus blossom or other deep tubular bloom, or to preen its feathers. Also, when avid with curiosity it hangs straight in front of your face, with vibrant wings flashing back in the sunshine the loveliest colours—metallic and iridescent. The emerald and bronze-green of breast and back, the dusky violet of the wings, the blue-black of the under-tail coverts, the black of the slightly forked tail of which the outmost but

·one feather on either side is greatly elongated and frilled on the inner margin, the coral-red of the bill with its black tip, all combine to make this tiny creature 'a thing of beauty and a joy forever'.

Five of these minute creatures were given to me by my Jamaican friends and two pairs of crested quail doves or mountain witch doves (*Geotrygon versicolor*). This latter is also becoming comparatively rare in its native habitat. The Reverend Dr. Gosse, in his famous nineteenth-century work, *Birds of Jamaica*, felt that: 'No description can give an adequate notion of the lustrous radiance of this most lovely bird.' Indeed, he was almost tempted to give it pride of place over the hummingbirds. The mountain witch, however, has much more subtle colouring than the gleaming hummingbirds, being of an overall bronze hue, shading to gold. Found more often on the ground than in the air it moves with great speed, merging in so well with its surroundings that, at first glance, it could be taken for a bundle of fallen leaves swirling along in the wind.

We have had a considerable success in breeding from those original birds also and the chicks have found homes in zoos in places as far apart as Hong Kong, Singapore, South Africa and Switzerland. The majority have been returned to continue breeding in Jamaica and a few are in zoos in this country. I was delighted to be able to send two pairs of mountain witch doves to East Berlin as well, as it seemed to represent the dove of peace and played some small part in cementing international relations.

All too often, however, such personal service is not available and many birds, imported in good faith, face such terrifying journeys that large numbers are dead on arrival or in such a poor state of health that they never recover. In an attempt to see just what was wrong with the methods of transport used I travelled on a freight plane carrying a cargo of birds and discovered that many of these tiny tropical species deteriorated rapidly when subjected to sudden changes of temperature, especially the intense cold over the Alps. To keep them as warm as possible I placed some birds under my shirt, directly next to my skin, with one under each armpit, a far more comfortable location for them than in the over-crowded luggage compartment of a possibly old and poorly pressurised aircraft. It was a successful experiment, although a messy one from my point of view, as the birds were not 'house-trained' and I ended up looking and smelling quite a rarity myself! Other ornithologists and animal lovers have campaigned for better conditions for transporting live birds and at least one airline has made improvements, although much remains to be done.

A cinnamon-breasted bee-eater perches happily on the hand of the photographer

However, I must confess that I too have been on the wrong side of the law when bringing back birds. It happened on that same return trip from Jamaica when I had been given the doves and hummingbirds. As I knew that great care would be needed to ensure that they survived the long journey in good health I particularly did not want to trust the minute hummingbirds to air freight, although I knew it to be illegal to carry live cargo in the passenger compartment of an airliner.

Prepared to take the risk, I made accommodation for the birds in a converted cigar box, with holes cut in the side and covered with a fine mesh of perforated zinc for ventilation. The compartments in which the individual cigars had been were left intact and the birds, wrapped in little cotton jackets for warmth (see illustrations) fitted snugly in each cubicle. As

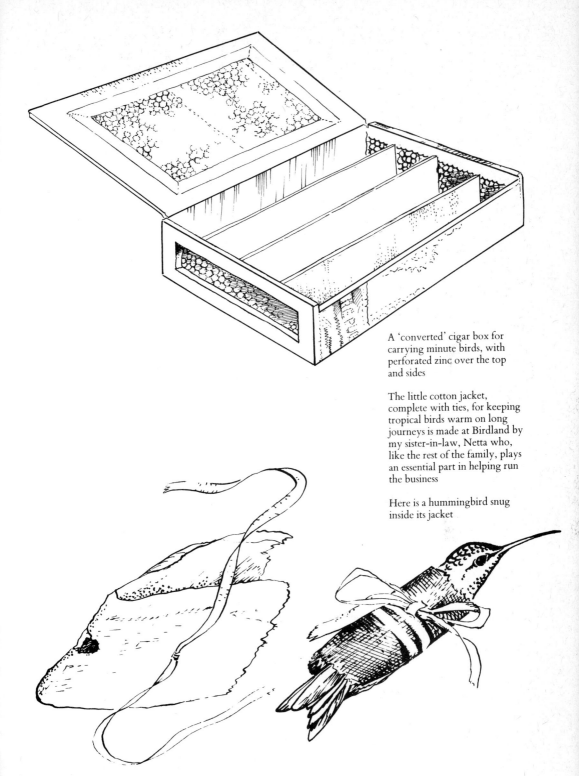

A 'converted' cigar box for
carrying minute birds, with
perforated zinc over the top
and sides

The little cotton jacket,
complete with ties, for keeping
tropical birds warm on long
journeys is made at Birdland by
my sister-in-law, Netta who,
like the rest of the family, plays
an essential part in helping run
the business

Here is a hummingbird snug
inside its jacket

(Opposite) King of the castle!
This hummingbird shows a
scant respect for its human perch

hummingbirds need nourishment every half an hour for twelve hours of the day, I made up several tubes of nectar, placed these in my holdall with the cigar box, then simply paid half-hourly visits to the toilet where I was able to feed the birds without fear of discovery.

The plan worked well on all the short hops up the Caribbean to the United States, and at Kennedy airport I took advantage of a two-hour wait to fill up the nectar tubes again. Hoping to find more privacy during the long haul over the Atlantic I changed my tourist ticket to first class, securing the last available seat. For an extra £80 I felt I had ensured security for my precious charges, as I knew that a sympathetic passenger sitting next to me would not object to my feeding the birds on my lap, instead of those somewhat embarrassing half-hour trips to the toilet.

Once airborne I relaxed and began chatting with my neighbour in the hope of testing out his reactions to my scheme. He failed to be drawn at first, expressing little interest in me and discussing the new Jumbo jet he'd just seen at the Lockheed factory. As he appeared to have such a strong interest in aviation I asked what he did for a living. 'I'm a director of BOAC', he explained, 'and my friend at the back is the inter-allied air transport manager.' My choice of companion could not have been more unfortunate. Fancy paying an extra £80 just to be landed with him!

Frantically searching for a solution to this new problem, I first asked if he would like the window seat, which would at least have prevented my falling over his feet on my regular thirty-minute excursion. As this was politely refused I then suggested that his friend might like to join him, but this idea too was airily dismissed with a 'Good heavens no, I've had his company all week and its much more interesting talking to you.' Just how interesting he had yet to discover. . . .

By now the first feeding time was nearly due and I had almost given up hope as he still went on making conversation. We discussed where we both lived and, when I mentioned the Cotswolds, my companion remarked that he and his wife were very fond of that part of the country, frequently staying in a hotel at a place called Bourton-on-the-Water. It seemed a happy coincidence that my home was in the same village. 'Oh yes', he continued, 'we know Bourton very well. In fact, we booked in there only recently then had to cancel it because of an outbreak of foot and mouth disease at home. We spent an hour last time in a marvellous garden full of exotic birds and we're determined to go back and spend longer there. I suppose you know it?'

'I expect you mean this place', I replied, reaching into my bag

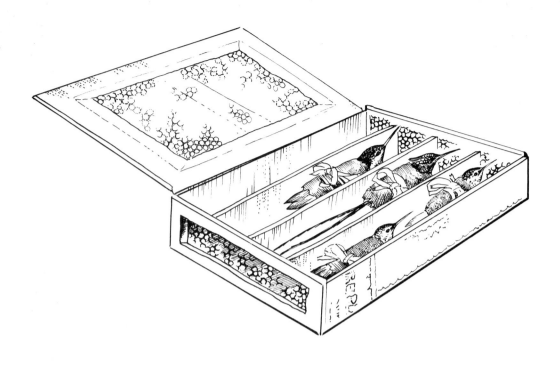

Showing how the
hummingbirds fit inside the
compartments of the cigar box
without any further conversion

for a copy of an aviculturalist magazine containing an article on
Birdland, which my wife had sent on to me. 'As a matter of
fact, it belongs to me.' Of course, he was delighted to make the
acquaintance of the garden's owner, and in such strange
circumstances too, but I had still not admitted my crime, and by
now the little birds must have been very hungry indeed.

'Before we go any farther, sir,' I said, reaching again in my
bag, 'I've got an admission to make. I'm afraid that there are
five extra passengers on this flight, without tickets—and here
they are', opening up the cigar box and starting to feed the
hummingbirds. Well, I've seen many surprised faces, but on this
occasion his eyes very nearly did pop out of his head.
Fortunately he was as thrilled as I was with the birds, ordered
champagne to celebrate and, at the end of the journey, said that
it had been his shortest flight ever over the Atlantic.

Heathrow airport, the day before Christmas Eve, was
extremely busy. My luggage was overweight and I was
bringing many birds and plants into the country. Piloting me
through customs without any difficulty my new friend offered,
at a later date, to include shots of myself and the humming birds
in one of his company's films. He was as good as his word and, a
few weeks later, we re-enacted that trip while the cameras

rolled and fire engines stood ready on the tarmac. Sequences from this film have now been included in one of my own films which has been shown all over the world. The Jamaican government included stills from it as part of their promotional campaign for tourism. The airline, of course, used it as still further proof to back up their advertising slogan 'BOAC takes good care of you'.

Caught in the act! My new-found friend from BOAC turns a blind eye to my illegal live cargo—hummingbirds travelling first class in a cigar box in my holdall

Jungle in the Cotswolds

Tropical species such as the hummingbirds can only survive in an environment closely allied to their native one. Always keen to have some of these exotic creatures at Birdland, I first converted an old greenhouse to form a small tropical house and this original building is still in existence, providing a perfect micro-climate for many small birds such as finches and the diverse family of contingas. Little African ducks swim in the pools and dart among the luxurious foliage, whilst radiant sunbirds can be spied next to the bright plants.

The uninitiated often confuse the magnificently coloured sunbirds with hummingbirds, but the two come from quite separate families as their movements reveal, the darting hummingbirds being relatives of the swifts and the sunbirds (*nectariniidae*) being perching birds. The 104 known species of this bird are restricted to the Old World, and are particularly numerous in Africa and India. We find it difficult to keep more than one at a time as they are extremely belligerent birds, fighting among themselves for territorial rights.

As I became more fascinated by the birds of the tropics it became clear that we would soon run out of room to house them all. A second, larger tropical house was the obvious answer, but I refused to be rushed into constructing the wrong sort of aviary, concentrating instead on building up a collection of tropical plants with which the birds would soon feel at home. I've mentioned previously that I'm very interested in plants and have been able to bring back unusual species from my excursions overseas. I must count myself lucky too in having sympathetic friends, such as the Minister for Agriculture in Jamaica with whom I stayed on my previously mentioned visit there. He made up a large bunch of anthuriums as a gift for my wife and, when we unwrapped them, we discovered a host of tiny plants hidden among the flowers.

This incident reminds me of the story of the keen American ornithologist who was anxious to breed birds in the United States from eggs obtained in Britain but was at a loss as to how he could get the eggs into his country as such traffic is against the law. Taking his courage in both hands he came up with a lightning, and apparently plausible, untruth: declaring the eggs, he avowed that they were part of a special diet which he had to follow for health reasons, and amazingly he got away with it.

(*Above right*) A group of visitors enjoying the flowers in the sunshine around the pool

(*Below right*) These jacanas, or lily trotters, live up to their common name as they step delicately on the water lilies

Whether he ever bred successfully from those eggs I do not know!

As the many tropical plants grew stronger, ready for the day when they could be transplanted to their permanent home, we took the opportunity to work on the existing garden which benefited from the past expertise of orchid-grower George Moore, although parts of it had still resembled a wilderness when we first opened. In those early days I was doubly fortunate in that a previous neighbour of mine had been the late Clarence Elliott, one of the grand old men of English horticultural writing, who was always prepared to give me advice and encouragement. When I first met Elliott I was living in a converted barn and he was convinced that it would be impossible to make any kind of garden from the poor surrounding soil, but as soon as he saw my method of going about the task we became firm friends and I learned much from him.

It might not be out of place here to say a few words about Clarence who was one of the bravest and funniest men I have ever known. At the time of our first meeting he was suffering from cancer of the throat. Despite being in constant pain and enduring many difficult operations he never gave in or lost his sense of humour. As he was unable to speak he fashioned a little man from a piece of wood and placed this figure on top of the dividing wall between our two houses. When he wished to gain my attention he would turn the man so that his face was towards me and, when I wanted to speak to him I would turn the figure towards Clarence's house. At all other times the little man faced straight down the wall.

With amazing resilience Clarence Elliott conquered that initial cancer, living for several years afterwards and regaining the use of his vocal chords. I shall never forget one occasion when he was still unable to speak and we were dining out at the end of a fishing holiday. I ordered for both of us and, as the courses progressed, the over-zealous waiter whipped away the menu each time before Clarence had time to study it properly. Eventually he could stand no more and, producing a large safety pin from his coat pocket, firmly anchored the menu to the table cloth. Back came the waiter, heaved at the card and took with it tablecloth, cutlery, glasses and all! Amused as I was by his discomfort I was only too glad to beat a hasty retreat from the restaurant before we were thrown out. Clarence, however, remained unrepentant, and through the post the next day I received that very same menu which he had taken as a souvenir.

Never one to let his long spells in hospital get him down, Clarence was always thinking up new tricks to play on the nurses. One day, when he had apparently had enough of tests

(*Opposite*) Some of Birdland's flamingoes in the dell where they love to hide away from the more public area of the lawn in front of the house

53

and examinations, he asked me to bring a ship in a bottle from his mantelpiece on my next visit. This I willingly did, and the next morning he produced it from underneath the covers, together with a note pronouncing, 'This is a ship which I passed in the night', much to his delight and the young nurse's confusion.

Clarence was a great sporting companion and we had many happy fishing trips together. Like me, he believed in doing jobs properly, and I well remember one night in 1960 when he went to Wales to collect carp for the garden pools. With three rods and an electric buzzer to indicate a bite, we primed our lines with large pieces of bread and cheese in individual one-and-a-half-inch circles. We fished marginally, catching the carp at the very edge of the lake, and in no time at all we were rewarded with a loud slurping sound as the fish rose to the bait. In all we caught fifteen carp that night, the heaviest weighing $14\frac{1}{2}$lb. Carrying them carefully to the car on wet sacks we quickly placed them in a galvanised tank full of lake water. We had also taken the precaution of borrowing a bottle of oxygen from our local dentist and, throughout the long journey home, we kept giving the fish short bursts. In this way all the carp were alive at

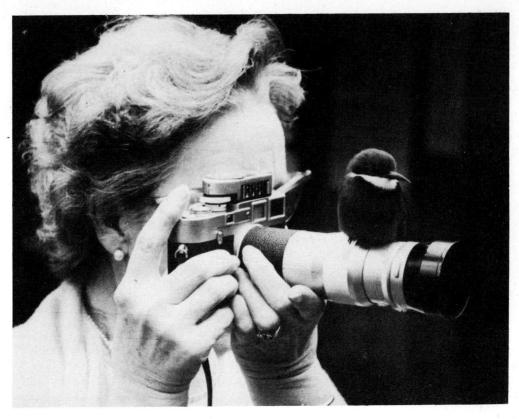

the end of the trip and I'm delighted to report that they are all still living today, some of them at least 100 years old.

This intrepid bee-eater seems to prefer being behind the camera to having its photograph taken

That there was a serious side to Clarence Elliott is revealed in his writings; his regular column 'In an English Garden,' which appeared in *The Illustrated London News*, is still looked upon as some of the finest, most forceful and idiosyncratic writing of its kind. Visiting Birdland in the early days he was good enough to describe me as 'a man with ideas, and the vision to convert ideas into reality, to reclaim the jungle'. He was particularly impressed with our specimen of the Californian gooseberry (*Ribes speciosus*), a leafy shrub, growing some six to seven feet tall with vivid scarlet fuchsia-like blossoms. And, he wrote too, 'It was grand to see a magnificent *Viburnum fragrans*, the finest and largest I have ever met, over 10 feet tall.'

As the main garden continued to flourish so too did the tropical plants and, after three years, I felt that the time had come to replant them in permanent surroundings. The new tropical house was to be built onto one end of Chardwar and I approached a famous construction company for a plan. They came very properly for consultation, then supplied blueprints for Victorian greenhouses with ventilators at the top, hardly the ideal place for keeping birds! So we came up with our own specifications and built the tropical house ourselves in half the time quoted by the prestigious company, and for half the cost.

Now one of the most popular areas of Birdland, the large tropical house is 60ft long by 40ft wide, 16ft high in the centre. A raised portion at one end houses the heating equipment and the water pumps. Here a waterfall has been constructed, running into a small pond where waders are often to be seen. The temperature is kept between 58° and 78°F (14.5° and 25.5°C) to ensure that the tropical birds never catch chills. This slightly steamy atmosphere, combined with the sudden, sharp bird cries, the pungent smells and the lush, dripping vegetation, all adds up to a feeling that this could easily be some part of a South American or West Indian forest.

The exotic plants with their bright flowers, unusual shapes and heavy perfumes back up this impression. The tropical house is designed in the shape of a miniature garden with a stone balcony serving as a barrier between the visitors and the foliage. Here can be found banana and pineapple trees in bloom (see page 123), and there is even a coconut tree which we grew from a nut. The luxuriant purple and red bougainvillaea flowers freely, as does the tropical American kiss-me-quick shrub (*Brunfelsia latifolia*) with its broad, leathery leaves and pale violet flowers, changing to white, with a white eye.

We grow four species of datura, all different but all equally attractive. *Datura cornigera*, sometimes called the moon flower, perhaps because of its creamy-white pendant flowers or because it is particularly fragrant at night, is a soft, downy shrub which hails from Mexico and grows to around ten feet high (see page 105). Beware of sinking into a soporific trance whilst admiring this plant as an ancient legend tells that the powerful aroma of the moon flower can put people to sleep, forever!

The white, funnel-shaped flowers of *D. candida* are responsible for its common name of angel's trumpet. This tree-like shrub comes originally from Brazil. Another trumpet-shaped flower belonging to this family with a rather more showy bloom is *D. mollis*, whose vivid, salmon-pink blooms suffused with orange can be found flowering freely in their native Ecuador. And finally we have the fragrant, yellow-flowered *D. chlorantha* from Peru (see page 105).

All my family have their favourite plants although it is difficult to select any one from such a profusion of glorious blooms. One of my personal choices would be *Clerodendrum thomsonae* (*balfouri*) from West Africa. Known as the glorybower, for obvious reasons, this twining evergreen shrub with glossy leaves and large, showy, pure white to pink flowers in clusters, climbs to a height of fifteen feet or more. Most

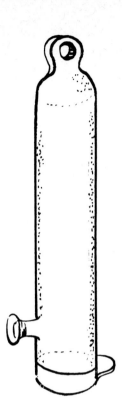

An example of one of our plastic feeders for hummingbirds which are now distributed throughout the world

visitors gaze in awe at the Malayan *Zingiber spectabile* which carries yellowish-white flowers in loose cylindrical spikes eight to twelve inches long, while others are impressed by the carmine-red flowers of the gorgeous evergreen shrub, *Medinilla magnifica* which comes from the Philippines and Java.

These, and other plants, such as the Mexican coral or fountain plant, *Russelia equisetiformis*, or the hairy tree, *Paulownia imperialis*, are grown not for their beauty alone, but rather for their usefulness in creating a compatible environment for the tropical birds, most of which rely on the flowers to provide them with nectar. Yet it is obvious that, with around thirty-four different species of birds in the tropical house, other forms of nourishment have to be provided. This is particularly so in the case of the deceptively fragile hummingbirds which eat twice their weight in a day. Darting about from plant to plant these tiny creatures gain some food from the flowers and water from the leaves but this has to be supplemented by home-made nectar which is now dispensed in our own special feeders (see illustration).

In the first, small tropical house we used glass phials which were difficult to keep clean as they were always breaking when

A hummingbird feeding on the wing from our own patented nectar dispenser

washed and thus proved expensive. More important, they were also dangerous as the birds frequently knocked off the glass teats, exposing themselves to possible cuts. In an effort to get round this I invented a plastic feeding tube which has proved so effective that we now export the design to zoos all over the world, thousands of feeders being supplied to hummingbird keepers in the United States alone.

The nectar in the tubes is made up of grape-sugar and a special American brand of invalid food containing all the necessary vitamins, a tablespoon of this being mixed with a quart of water and six tablespoons of sugar. The birds consume about ten pounds of sugar a week, so naturally I was a little worried during the 1974 sugar crisis in Britain when stocks over the counter were limited. I therefore wrote to the major manufacturer, stating my predicament and asking if I could obtain sugar direct from the company. Back came an immediate reply, stating that certainly I could have the sugar, but expressing concern about my capacity to store it in bulk as it was supplied only in twenty ton lots! They were able, however, to suggest a wholesaler who provided bulk sugar in five-hundredweight bags, so we quickly sterilised five dustbins and kept our load in these when it was duly delivered. Grateful that my birds were not missing any food I was careful not to tell any local housewives that the refineries were obviously full of sugar for the asking—plus, of course, a small consideration.

Hummingbirds need so much food as they generate an enormous amount of energy from it. In comparison with the rest of their body size, these tiny creatures have the largest heart of any living thing, giving the highest energy output per unit of weight of any known warm-blooded animal. When introducing a BBC TV 'World Zoos' programme from Birdland—incidentally, the smallest zoo ever to be included in the series—that eminent naturalist, the late James Fisher, worked out that from their daily sugar ration, the birds produce the equivalent of 2,000 horse power. And, as Crawford H. Greenewalt states in his beautiful book *Hummingbirds*, when hovering, a hummingbird has the energy output per unit of ten times a man running at nine miles an hour. Greenewalt also points out that if a 12 stone (170lb) man in a fairly active job has an energy output of 3,500 calories then a hummingbird's equivalent is 155,000 calories. Staggering isn't it?

The majority of this energy is spent in the sheer act of flying. Hummingbirds gain their name from the hum of their wings as they beat through the air producing a definite high-pitched sound. Some of the larger varieties, like the comparatively well-named giant hummingbird, which is about the size of a swallow, have wing beats of fifty to the second, but the tiny Cuban bee hummingbird, for instance, with a half-inch body (you get fourteen birds to the ounce) makes as many as one hundred wing beats per second. The larger the bird, the slower the beat.

An incredibly fast mover, the hummingbird can fly backwards with ease. Rather like a helicopter movement, hummingbird wings beat not only up and down but also backwards and forwards so that there is a flow of air downwards. The birds' wings cannot rotate, however, and must be rearranged at the shoulder for every forward and backward movement. It is impossible to film such movement without the aid of special equipment as the speed is much too fast for the camera to follow, but my German photographer friend, Franz Lazi, some of whose superb colour photographs illustrate this book, has captured the hovering flight of the hummingbird feeding in our 'Birdland Story' film, a still from which is shown on page 70.

Many ornithologists proclaim the hummingbirds among the most beautiful of birds, and certainly they are stunning with

Fearless little hummingbirds
fluttering round a visitor's head
in the hope of picking out a
hair or two for their nests

their iridescent colours which change in flight as shafts of light
catch their feathers. Among the many colours, greens
predominate, but there are also reds, blues, yellows and purples,
many of the colours being reflected in the birds' names, like the
ruby-throated hummingbird of America. Sadly, the birds'
voices do not live up to their looks as they have no definable cry
apart from the constant twittering amongst themselves.

Probably because they are such proficient flyers, even bathing
on the wing and taking their food while in motion, the
hummingbirds seem to have lost the use of their legs, if indeed
they ever knew how to use them. Sidling on a perch, they
simply cannot walk and immediately take to the air when
disturbed, rising straight upwards as if rocket-propelled. Such

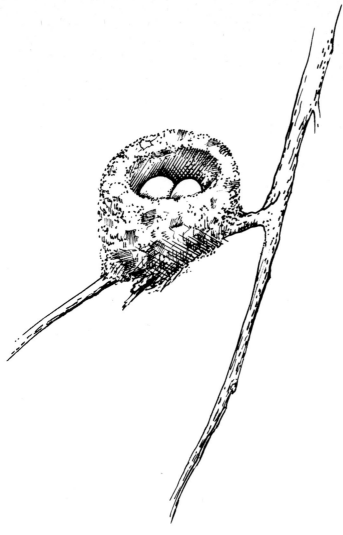

constant demands on their energy must be physically exhausting but the ingenious hummingbird has a metabolism well designed to cope with this. When sleeping, the little birds have the facility to go into a trance-like state of suspended animation, during which time their body temperature drops drastically and their energy output is extremely low. This does not necessarily occur every night, although the birds seem capable of attaining this state of near coma whenever they wish, a state so pronounced that they can be handled without appearing to notice.

In the daytime, however, it is quite a different story as the hummingbirds show a complete disregard for the presence of humans and alight on visitors without any show of fear.

This nest of the vervain hummingbird is stuck together with the bird's saliva. They also will take human hair and dip it into the neck of the nectar dispenser to make it sticky so that it helps glue together the other materials

Curious, fearless and pugnacious when defending their own, they constitute no threat to people, frequently taking a strand from a mohair sweater for nest-building, or even picking a human hair from a passing head! As can be seen from the drawings, the hummingbirds' tiny, cocoon-like nests are miniature construction masterpieces but, unfortunately, such is their appeal, that some of our more unscrupulous visitors cannot resist taking the nests home for souvenirs. We even lose a few birds themselves in this way. It is impossible to keep a check on all these tiny creatures and we have discovered that one or two have been smuggled out. Thankfully, this only happens rarely, and on the whole there is a minimum of theft and vandalism in the garden.

It would be difficult for even the most determined thief to hide away one of the most distinctive of our tropical birds, the brightly coloured cock of the rock, now housed in a new aviary. A native of Bolivia, South America, this bird of brilliant plumage is extremely rare and little is known about its habits. Fruit eaters, particularly fond, we find, of grapes, these medium-large birds are about twelve inches long with a solid appearance. It is possible to tell the birds' countries of origin from their colour, a bright, orange-yellow denoting natives of Brazil and Guyana, whereas the flame-red birds with black wings come from Colombia, Peru and Bolivia.

The mating antics of the cock of the rock are worthy of record, with the male performing an extraordinary dance routine to attract the smaller, less brightly coloured females. In the wild, the males assemble before the females in a clearing in the forest and each, in turn, goes through an intricate act, strutting and flapping to prove that they are worthy of their chosen mate.

If prizes are being handed out for colouring then the resplendant quetzal (see page 87), one of the *Trogon* family, must be in with a good chance. Like a giant hummingbird, it is predominantly a vivid green with extremely long tail feathers so that it measures some four feet in all. Its plumage is so weakly attached to its fragile skin that body feathers will fall out unless extreme care is taken in handling. A bird of ancient origins, the quetzal was worshipped by the Aztecs as a symbol of freedom and its name is given to Guatemalan currency. Today it has pride of place on the Guatemalan national flag. Its call, a low double 'coo', is deceptive as, like a ventriloquist, it seems to throw its voice, so that although it can be heard in one place it cannot be seen there.

Finally, it would not be permissible to even mention beautiful species without reference to the birds of paradise (*Paradisaeidae*) whose fantastically coloured plumage, often displayed in bizarre

patterns, is an unforgettable sight. I had been trying to obtain some of these marvellous birds for some years but was never lucky enough to find any and also did not have any suitable accommodation. Then, by chance, I was offered some birds of paradise and, at the same time, was fortunate in that one of my neighbours retired so that I could purchase his land and construct a special home for these newcomers. After several months' acclimatisation the birds were ready to be put on show and we had a nice little get-together when Johnny Morris kindly came along and officially pronounced the birds of paradise aviary well and truly open.

I'd like to end this chapter on a more sombre note by repeating some of the remarks made by Johnny on that occasion. He pointed out the fate that is befalling many of our river and underwater life through mindless pollution and greed. The 10,000 square miles of Lake Erie in North America, for example, are practically dead, devoid of fish and vegetation. Thankfully, laws have been passed making the slaughtering of hummingbirds for the millinery trade illegal. Hopefully, too, the growing repugnance to the wearing of the skins of rare wild animals will continue to grow. But we must be ever on our guard that the opening up of new territories will not lead to further wholesale slaughter of all those harmless wild creatures which can offer no defence against gunshot.

King of the Penguins

Although all our birds are regarded as members of one big family it is impossible to get to know all the inhabitants on a personal level, and I suppose that I may be accused of 'favouring' the more communicative members such as the parrots and macaws. One particular group of birds have a special place in my affections, however. I'm referring, of course, to the penguins. A small colony of these delightful creatures was established early at Birdland and has continued to flourish over the years. Indeed, so successful have been our methods of raising these birds that some of them have now reached the grand old age of sixteen and are still going strong. The life expectancy of a penguin in the wild can be up to thirty years but, sadly, the average life-span of these birds in captivity is still only about two years.

Perhaps we are simply lucky, but I firmly believe also that our penguins are happy and healthy because we try to provide an environment as close to their natural one as possible. With our variable climate, it is impossible to achieve the pure air of the Antarctic which prevents the breeding of bacteria, but over the years we have had a remarkably disease-free record, a fact which led James Fisher, in his *World Zoos* book, to comment that the Birdland collection of penguins must be one of the finest in captivity.

Bourton-on-the-Water is situated 480 feet up in the Cotswolds and Birdland is fortuitously placed over a subterranean lake only ten feet below in a gravel strata. This means that we have a constant supply of fresh water on which to draw, an important factor in helping us to provide the correct facilities for the penguins. Two electric pumps working day and night, circulate water at a constant 46°F (7.7°C) through the penguin pools, ensuring cleanliness. The site of the penguin colony has a northern aspect with the large pool never receiving direct sunlight. This position was deliberately chosen to simulate Antarctic weather conditions for, although penguins in the wild are quite accustomed to sun, they prefer shade, and the customary cloudy, misty weather in the Antarctic provides this.

An old coaching house on the site provided a perfect diving tank for the penguins when we converted it by removing the doors and constructing a false portcullis to form an archway.

King penguins out to take
the air. The little rockhoppers
in the background seem to know
who s boss

The tank has a one-and-a-half-inch-thick plate glass window at the front, acting as a viewing chamber where visitors spend hours watching the agile birds perform their sub-aqua feats. Water cascades from the tank into a smaller pool which drops away from a shallow pebble beach. This latter might be thought of as an unexpected luxury but it is a curious fact that pebbles are an aid to the penguins' digestion. King penguins, for instance, can frequently be seen swallowing several pebbles, often as big as walnuts.

Standing over three feet high, the large kings, with their metallic grey coats and snowy-white plumage are amongst the garden's major attractions. These striking birds are easily recognised by the large, comma-shaped yellow ear patch which extends as a narrow line to join a golden patch on the breast (see

(*Above right*) The tiny Hahn's or red-shouldered macaw which I have successfully bred and returned to its native Trinidad

(*Below right*) It is easy to see how the streamertail hummingbird got its name, and also its nickname of the doctor hummingbird from the black plumage reminiscent of the frock-coats worn by old-fashioned doctors

page 106). A distinctive iridescent shade of green on the crown and throat is a sign of good health. In young kings, the attractive splashes of colour are duller, the fledglings being covered in a thick, woolly down which causes them to look like cuddly brown bears.

When breeding, the king penguin is remarkably tame and is less disturbed by human intrusion than others of the species. It makes no nest but incubates the egg on its feet, covered by the tail behind and a loose fold of skin in front. It can walk, though awkwardly, and the casual observer would not know that the bird was incubating. This incubation period lasts for fifty-six days but is less with other kinds of penguins. We were desperately proud of the first king to be hatched here and it caused quite a stir in the press. Named Pinook by Sir Peter Scott's children, it is still in the garden today.

In all there are some seventeen varieties of penguins, most of which are found in the Antarctic regions but a few, such as the Humboldts, living within 3° of the Equator. We have a good selection at Birdland including, as well as the kings, Humboldts, the Magellanic or jackass penguins, rockhoppers and the gentoo. All live happily together and will only scrap amongst themselves for herrings and sprats tossed into the pools when

I wonder what this mother king penguin is saying to her ten-day-old offspring?

'official' feeding time comes round at 11.30 every morning, an event which the visitors, especially children, seem to enjoy almost as much as the penguins!

At other times, when we wish to give medicine and extra nutrients, the penguins are also fed by hand. As described in Chapter 2, their main diet is fish. This is transferred from the deep freeze on the day before use to a special 'cold' room which enables the fish to thaw out slowly. Each day's ration is taken from the room and put into a Belfast sink where water is sprayed over it till half an hour before feeding time. Then all the damaged specimens are sorted out as it is imperative that the penguins are fed only whole fish. A broken bone could easily score their throats, setting up inflammation which leads to a fatal disease known as aspergillosis, in which a fungus builds up in the lungs and eventually prevents breathing. This harsh disease, much prevalent in the wild, is impossible to treat and causes the penguins to die in two to three days. Disastrous as this may be in captivity, in the birds' breeding grounds it is looked upon only as part of the natural cycle in which penguin carcases provide food for the scavenging varieties such as petrels and skuas, so ensuring a balance of species.

I have always been fascinated by the idea of seeing penguins on their home ground, so was particularly delighted to be asked to join Peter Scott on his 1968 expedition to the Antarctic. Following partially in the footsteps of his famous but ill-fated father, Peter was hosting what can only be described as the first tourist trip to the southernmost point of the globe, although the majority of us were naturalists and he personally was using the visit to study rockhopper penguins in their natural habitat. We flew to South America and, in the civilised atmosphere of Buenos Aires, found it somewhat difficult to believe that we were only 1,000 miles from the eternal ice, the virtually uncharted 11 million square miles which lie south of the 60° latitude. Yet, as we flew over Patagonia, close by Cape Horn, the stormiest place on earth, and saw the black, infertile soil still bearing traces of the Ice Age, we could well imagine the days when the ice sheet used to reach the Argentine. And, although subduing to us all, pampered by western sophistication, it seemed strangely appropriate to hear that the average life expectancy of the remaining native Indians of this wild region is still only thirty years.

Our first major port of call was Stanley, the capital of the Falkland Islands, which lie 200 miles from Patagonia, being composed of two large and thirty smaller islands. Belonging to the British since 1833, the Falklands have always been a bone of contention between successive UK governments and those of the Argentine. There are approximately 2,400 inhabitants on the

(*Opposite*) Technically a very difficult photograph to take, this shot shows a Jacobean hummingbird taking nectar from an *abutilon* whilst on the wing. Note the blur of the rapidly beating wings

(*Below*) The head of the king penguin is distinguished by the large, comma-shaped ear patch

(*Right*) The gentoo penguin is distinguished from other varieties by a white bar over the head

(*Below left*) The rockhopper penguin with its brilliant yellow eyebrows which end in tufts

(*Below right*) The head of the macaroni penguin, also to be seen at Birdland

islands, who have little knowledge of their British forebears and are now employed mainly in sheep rearing.

One of the highlights of the venture for me was a meeting at Stanley with the Reverend Millan who had been responsible eight years earlier for purchasing Birdland's first king penguins from South Georgia, acting as my agent and arranging transport. As about £140 was left over from the £1,200 which I had sent him for the transaction, I had suggested that he put the majority of this sum towards the installation of a central heating system in his deanery and that the rest would buy a ceremonial cope to wear on special occasions in the cathedral. I was deeply touched that he considered our visit such an occasion and put on the cope for a service which we attended.

Throughout our stay we were most hospitably received by all the people of Stanley including 'Dick' Barton, a local magistrate, with whom I struck up a real friendship. But in next to no time it seemed we had boarded our ship for the voyage to the tip of the Antarctic, and the welcoming populace of Stanley soon paled into insignificance beside awe-inspiring Antarctic thunderstorms, gigantic icebergs of which it was difficult to believe that only one-seventh was visible, and the breeding grounds of five million penguins which feed annually on 9,000 tons of fish and crabs. Like Coleridge's *Ancient Mariner*, our ship was always pursued by the wandering albatross, one of the largest birds in the world with a wing span of ten feet. Seals, dolphin-like, leaped out of the water as they followed our progress and we passed floes carrying seals and sea lions.

The glaciers mark the beginning of the seventh continent where only a few hundred men live in research bases. But, as man diminishes, so the wildlife flourishes and, the farther south we pressed, the tamer became the animals. The baby seals met us

(*Below left*) Humboldt penguins can be found living within 3° of the Equator

(*Below right*) Easily distinguished by the pattern of black and white bands crossing from head to chest, the Magellanic penguins are commonly known as jackass penguins because of their donkey-like braying call

This group of penguins seems to appreciate the fish fed them by Sir Peter Scott with puppy-like zeal, an enthusiasm which they must have shown to the nineteenth-century seal farmers who slaughtered these creatures in their thousands for their pelts. Now, thankfully, there is a closed season for seal hunting which prevents extermination of this lovable mammal, clumsy on land but perfectly co-ordinated in water.

One of the most memorable sights of that trip, and indeed of my entire life, was at the end of a day, looking out over a bay, when we saw at least 10,000 Wilson's petrel feeding. These small, black birds looked like a dark cloud, so many were their numbers. Another day, we arrived at the island of Deception after a volcanic eruption, when the sea was still boiling with sulphur. Here we saw bleached whalebones and the decayed remains of boats, reminders of the horrors of the early decades of this century when the whale was slaughtered *en masse*, its oil being processed for soap, the thin plate from its upper jaw for corsets and the ambergris for perfume. In one of these bays, in 1941, two German blockade-runners were captured by the British fleet. Since that time the penguins have remained undisturbed and the absolutely fearless chinstrap penguins will allow humans to approach extremely close so long as no sudden movements upset them.

It seems remarkable to contemplate that when the earth was middle-aged penguins could fly. Now they are almost pure water birds, picking up speed under water as they approach land and leaping ashore, taking advantage of the breaking waves. With crops filled with fish and crabs they waddle up the shore to satisfy the hunger of their quarrelling offspring. We were able to observe at first-hand the very pronounced social structure of penguin colonies, living as they do in more or less permanent pairs, producing only one baby at a time and fiercely guarding the egg when incubating. Some of the old birds are employed as nursemaids of the young, but even so the infant mortality rate is very high, up to 80 per cent, as the evil-looking skuas, known as the eagles of the Antarctic with their hooked beaks, are always searching for unguarded eggs or lonely young.

The Antarctic latitudes are known to sailors as the 'roaring forties, howling fifties and screaming sixties'. Summer comes to an end in October and by mid-March only the routes to the research stations are kept free of ice. Temperatures range from 20°F (−6.7°C) in summer to −17°F (−27°C) in winter, a time when ice becomes a real danger to ships and the base workers begin to suffer from sheer ice monotony. Of course, we only touched the tip of the Antarctic but we were lucky enough to stop off at Palmer Station, one of the research camps where people from eleven nations were working peacefully together on joint projects which, as yet, have not born fruit in our crowded world, but may one day provide some answers to our food and survival problems.

Franz Lazi had accompanied me on the trip, making a film for German television and I was later able to obtain a copy of this for my own use. On my return we had a special film show for some of our friends and were delighted to have as chief guest the late Sir Raymond Priestley, then the only surviving member of Scott's original expedition. Seeing so many beautiful birds living free and unafraid had been an unforgettable experience for me, and one which I at first found difficult to put behind me until once again I became involved in the day-to-day running of Birdland.

Some months later, by that same amazing coincidence which seems to have been part of my life, I was a guest at a wedding when, spying a familiar face to which I could not, at first, give a name, I suddenly realised that it was my old friend from the Falklands, 'Dick' Barton. Naturally conversation centred on my visit and I got quite carried away extolling the virtues of the area as a wildlife sanctuary. 'Well, if you're so keen, why don't you buy your own island over there?' Bartie asked innocently. It turned out that he was executor to a will in which two of the Falkland Islands, Grand and Steeple Jason, were to be sold. The 10,000 acres of land on which 5,000 sheep grazed had already been advertised in the London *Daily Telegraph*, but so far there had not been any interest expressed. Tempted as I might have been by the thought of all those birds I did not relish the idea of taking up sheep farming and the purchase price was also beyond my means.

Once again I dismissed the Falklands from my mind and concentrated on the Cotswolds until, out of the blue, I received a letter from 'Bartie' offering me both islands, minus sheep, for £7,500. This was much more to my liking and my financial advisors were prepared to countenance such an 'investment' although we had no idea of how to go about effecting the purchase. Faced with the prospect of enlarging the Birdland business to include two rocks in the southern hemisphere, my accountant weakly suggested that I should consult a solicitor with Commonwealth experience. The islands, together with all the wildlife thereon, had been sold by the British government in 1861 to a company with the name of Dean Brothers and, as the land was being sold now by the wills of the owners' direct descendants, we suspected that there might be difficulty with the conveyancing. Was I to be disappointed at this late stage?

Taking the sheep, literally, by the horns, I booked a call to Stanley and made Bartie an offer for the islands of £5,500 without the sheep. Although this was below the previous asking price, as there were no other interested parties he accepted. I

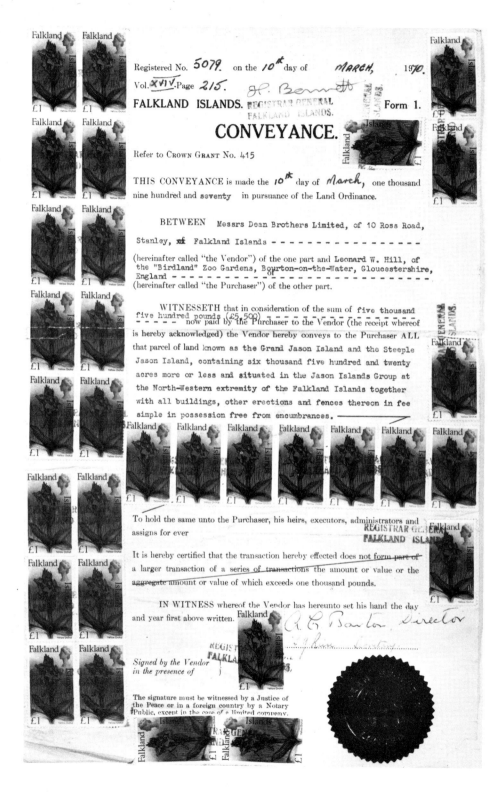

Registered No. 5079 on the 10th day of MARCH, 1970.

Vol. XVIV. Page 215. *J. P. Bennett*

FALKLAND ISLANDS. Form 1.

CONVEYANCE.

Refer to Crown Grant No. 415

THIS CONVEYANCE is made the 10th day of March, one thousand nine hundred and seventy in pursuance of the Land Ordinance.

BETWEEN Messrs Dean Brothers Limited, of 10 Ross Road, Stanley, of Falkland Islands - - - - - - - - - - - - - - (hereinafter called "the Vendor") of the one part and Leonard W. Hill, of the "Birdland" Zoo Gardens, Bourton-on-the-Water, Gloucestershire, England - - - - - - - - - - - of - - - - - - - - - - - - - (hereinafter called "the Purchaser") of the other part.

WITNESSETH that in consideration of the sum of five thousand five hundred pounds (£5,500) - now paid by the Purchaser to the Vendor (the receipt whereof is hereby acknowledged) the Vendor hereby conveys to the Purchaser ALL that parcel of land known as the Grand Jason Island and the Steeple Jason Island, containing six thousand five hundred and twenty acres more or less and situated in the Jason Islands Group at the North-Western extremity of the Falkland Islands together with all buildings, other erections and fences thereon in fee simple in possession free from encumbrances. ————

To hold the same unto the Purchaser, his heirs, executors, administrators and assigns for ever

It is hereby certified that the transaction hereby effected does not form part of a larger transaction of a series of transactions the amount or value or the aggregate amount or value of which exceeds one thousand pounds.

IN WITNESS whereof the Vendor has hereunto set his hand the day and year first above written.

R. G. Barton, Director

Signed by the Vendor in the presence of

The signature must be witnessed by a Justice of the Peace or in a foreign country by a Notary Public, except in the case of a limited company.

then explained our thoughts over the difficulties involved in actually making the payment, but he was able to reassure me that there should be no problems as this was a registered title. 'I'm coming to England in a month or so's time,' he went on. 'Why don't I pay the stamp duty on your behalf, pick up the land register certificate and simply bring the deeds over with me?' Such a suggestion was fine by me and I found it particularly satisfying that two men, nine thousand miles apart, could settle a major decision so speedily without undue legal fuss or complication.

Sure enough, all went well and the deed was duly delivered. It looked such a pretty piece of paper (see illustration) that I could scarcely believe that it gave me access to all those wonderful birds. Now I was truly 'king' of the penguins, though I had still not set foot on my islands.

(*Opposite*) The deed of conveyance which made me a real penguin millionaire, owner of Grand and Steeple Jason Islands

Len Crusoe

No land owner can have had a stranger introduction to his territory than on my first visit to Grand and Steeple Jason. The long haul to the Falklands is difficult enough from this country, being made by air, via Madrid, Sierra Leone, Caracas, Rio de Janeiro and Montevideo. At that time too there was no regular air service to the Falklands, although one has recently come into operation, and I had to endure another four-and-a-half days by ship to Stanley. From there transport to the Jasons is a very haphazard business as I was soon to discover; the unruly seas around the islands are a vast graveyard for hundreds of shipwrecks so few sailors venture forth without good reason. On that first trip I spent three days at Stanley in appalling weather and was beginning to wonder whether I would ever get a chance to see my islands, let alone set foot on them, when I was offered an air lift by the local postman, Ian Campbell, who regularly delivers mail to outlying communities in his little Beaver aircraft. I jumped at the opportunity to get away and was as thrilled as a schoolboy with the start to the proceedings when I played postman for the day, dropping the letters in padded bags from the low-flying aircraft to the tiny figures waiting below with outstretched arms.

When preparing to visit the islands I had arranged for Ian Strange, a naturalist and resident of the Falklands, to go out to Steeple Jason before my arrival to fix up living accommodation and rations. As our little plane approached the area we saw Ian waiting near the shore, but unfortunately the sea was too rough for us to land in that particular spot and we had to skirt the coast in search of calmer waters. We eventually found a suitable stretch of water and the aquaplane made an excellent touchdown, although it had to take off again as soon as I had unloaded all my belongings as there was a storm coming up. Waving farewell to postman Ian, I took up my luggage and began scrambling over the rocks to where I could meet up with Ian Strange. I soon stopped in even more haste than when I had started out. Ahead was not dry land but a twenty-foot deep channel some ten feet wide, separating me from the mainland. With a high wind blowing and the sea pounding below I was stuck, high and dry on a solitary rock with no sign of Ian!

As I pondered on my plight, curious 'Johnny rooks' swooped around me. The correct name for this carrion hawk is striated

Len Crusoe surrounded by
his subjects

caracara (*Phalcoboenus australis*) but it has a 'caw'-like sound, somewhat like the European rook which caused early settlers in the Falklands to give it the more familiar name. A big, dark bird, it is inquisitive by nature and remarkably tame. When Charles Darwin visited the Falklands in 1833 he reported that the striated caracara was 'exceedingly numerous' and that some were always waiting near the houses for scraps or offal. Today, however, the species is rare, being confined to the more remote islands such as the Jasons, as 'Johnny rooks' were persecuted after the introduction of sheep to the Falklands because they were reputed to kill young lambs.

Interested as I was at the chance to see such rare creatures at first-hand, I did not exactly relish the circumstances which had made it possible. Not that they would dream of regarding me as pickings but my state of mind was such that I found it extremely difficult to conjure up my old vision of me as a lord of the manor addressing his avian subjects, and indeed I began seriously to wonder whether my purchase of the islands had been a wise step after all.

Eventually, after an hour or so, Ian Strange's figure came into view, making his way across the slippery surface of the mainland. His smile of welcome soon turned to dismay as he realised my predicament and his extended arm was seen to be of little use across the divide. Not the person to be put off by such minor difficulties, he searched around and was able to find a long piece of rail from the old sheep-shearing days, which he hoisted over to my rock. A piece of fencing wire completed our salvage equipment and, by wrapping this firmly round the cases, we slid them, one by one, across the rail to safety. Now only I remained. Stepping out warily as far as I could along the rail, I then leapt for my life towards Ian's outstretched hand. I landed in water up to my middle, but this seemed little discomfort after the thoughts which had been going through my mind and, at last, I could say that I had arrived on my island!

Although the weather was fairly mild, the island presented rather a bleak prospect for those looking for conventionally beautiful wild scenery. Any ground which was not covered with birds or nests was rocky and bare apart from huge clumps of tufted grass. There was one mountain of any size with rocky outcrops on the more low-lying hilltops, from where could be obtained a view of the deeply indented coastline with boulder-strewn beaches. There are no native trees but the contrasting colours of the sand, the vegetation and above all the thousands of birds, made the island look especially appealing to me.

As we tramped through the rookeries Ian dropped another bombshell: the special boxes which I had prepared for transporting back wildlife had been left at Stanley. One of the main objects of my trip, as well as just an introductory visit to satisfy my curiosity, was to take some birds, penguins in particular, back to England. To this end I had designed collapsible packing cases, with hinged lids and air holes, which had been air freighted unassembled to the Falklands. The plan had been for Ian to take them with him when he went out to the Jasons to prepare for my arrival, but unfortunately they had not been loaded onto the *Forest*, the ship which transported him to the islands.

As we were to be taken off the island by the same ship in a few days' time, I realised that my journey need not be wasted if I could get a message to the mainland asking that the boxes be put on board. We had been supplied with a field radio by the marine department at Stanley to report back on our movements so, as soon as we had made ourselves as comfortable as possible in one of the old-shearing huts which was to be our 'hotel' during the visit, we fixed up the aerial with a piece of string and prepared to broadcast. Not a spark of life came from that machine. We waggled the aerial, we turned the handle, we shouted hello but only the seals answered us: the radio remained dead.

For three days we attempted to broadcast and for three days we failed to make contact. By this time all my forebodings were beginning to return. Beginning to panic we decided that reception might be better at the top of the mountain so, the next day, carrying all the cumbersome radio equipment, we scaled a good 1,200 feet, only to be greeted with the by now customary silence. The soft hues of the Cotswold countryside grew hourly more appealing. Perched on top of the mountain, after an almost superhuman effort on my part to get up there, Ian suddenly decided that the fault lay in the aerial which was not positioned in a damp enough spot for good reception. Off down the mountain with the aerial he charged in an effort to make contact before the airwaves ceased at 3.45pm, with me stumbling behind, carrying the heavy equipment. Once again the aerial was earthed, we prepared to broadcast and heard—nothing.

By this time tempers were beginning to get slightly frayed. Unable to sleep that night I was searching through some of the old boxes in the shed and, miraculously, came across the radio operating instructions. The fault was nothing to do with poor reception and the positioning of the aerial: we had simply not known how to work the damn thing! The next day, following the instructions to the letter, we made contact with the mainland to be greeted by relieved cries of 'Where the hell have you been these last three days?' Apparently a constant lookout had been maintained to watch for distress flares from us and

The somewhat forbidding
terrain of Steeple Jason Island
looking towards Grand Jason

(*Above left*) Three tiny carmine bee-eaters poised in the tropical house

(*Above right*) The brilliant green feathers of the long-tailed quetzal blend in with the bright tropical foliage

(*Below left*) Surely a contender for the title of most colourful bird in the garden, the rainbow lory from Australia

(*Below right*) But the scarlet cock of the rock with its amazing courtship dance might well argue the toss on this point

rescue parties were standing by to come and fetch us off the island, but we were able to pass on the message that we were well, if frustrated, and arranged for the ship to fetch us in three days' time.

Now that we knew when to expect the packing cases we could begin to collect the penguins, but once again we encountered a major hitch: with no boxes to hand we had nothing in which to keep the birds once they were caught. So, once again a makeshift solution had to be found. Scattered around the hut were fencing stakes and we spent a morning collecting these and then driving them into the ground in a 14ft circle. How we blessed those sheep farmers as we discovered wire netting left by them which we joined round the stakes to form a corral. It soon became obvious, however, that whilst this would contain the penguins for a brief period, as long as they could see out they would try to escape, become frustrated by not being able to do so, and would only succeed in harming themselves, mentally if not physically. Yet again we had cause to thank the previous tenants as we found bales of heavy old sacks which had once been used to contain wool and a ball of twine complete with a bagging needle. Ian proved an expert at tacking and we were thus able to encircle the netting with the sacks sewn together to create a suitably dark place.

If our operations had been difficult so far they began to resemble a scene from a Crazy Gang show when we set about actually catching the penguins. We had to be selective as I wanted only youngsters, born the previous year but not yet breeding. One other qualification was that we preferred birds just coming to moult as these would require less food and thus stand a better chance of surviving the long journey in good condition. A penguin in moult can survive for up to three weeks without food, so we were looking for plump young birds, going grey on the top of the head with feathers just rising on the breast, sure signs that the moult is about to begin.

I wanted to catch such birds without frightening their companions and creating general panic among the penguin colonies. Loath to pounce on the selected birds by force which could hardly fail to rouse the others I found yet another piece of stiff wire and bent this round to form a crook-like implement. Our method was then to slide the wire along the ground towards the desired penguin and, as it turned away, slip the hook over its foot and draw the bird briskly towards us along the ground, as shown in the illustration on page 92. This little device proved extremely successful and in three days we were able to catch sixty-six birds.

We took mainly rockhopper and gentoo penguins. The rockhopper (*Eudyptes crestatus*) is the smallest penguin breeding

in the falklands and possibly the most attractive, being blue-black above and white below with broad, brilliant yellow eyebrows which end in tufts of long yellow plumes behind the small red eyes. The name rockhopper, or 'rocky' as it is sometimes known, comes from its habit of scaling steep rock faces in energetic bounds, with both feet together. Generations of rockhoppers scrambling up the slopes of Grand and Steeple Jason have scored deep vertical grooves in the rocks with their claws. Thousands of them nest on the islands in close-packed rookeries during the breeding season between late September and April. The noise from such a colony is deafening at close quarters and you will gain some impression of the sound if I tell you that the call of the rockhopper has been described as being like the grinding shriek of a rusty wheelbarrow being pushed at full speed!

After breeding, these penguins desert the rookery and lead a wandering life at sea in the South Atlantic. Here, the rockhopper has two main enemies, the sea lion and the leopard seal, both of which capture swimming penguins. This is not a pretty sight as the seal, for instance, surfaces under the bird and flings it violently back and forth several times in order to split and tear off the skin before swallowing the carcase. Many people, I know, complain about the supposed 'cruelty' of keeping wild birds in captivity but such critics might stop and think of the constant battle for survival which represents the majority of lives in the wild, and many birds in good zoos often have far longer lives than if they had to fend for themselves in their natural habitat.

The gentoo penguin is also very numerous and widespread in the Falklands with colonies of several hundred pairs on both the Jasons, occupying low plateaux near the shore. Distinguished from other varieties of penguins by a white bar over the head and a long orange and black bill, the gentoo is not as tall as the king penguin. It will stand proudly and approach inquisitively if people remain still, but will panic and rush away if suddenly disturbed, toboganning down slopes on its breast.

The gentoo makes use of the native shrub, 'diddle-dee', in building its bulky nest in which to lay two large white eggs about the size of tennis balls. They tear the bushes up, piece by piece, so that soon after the eggs have hatched the ground is bare apart from the nests. Local sheep farmers on the other islands look favourably on the gentoos as the birds' droppings encourage the growth of new grass after the diddle-dee has been killed, and the sites of colonies are changed slightly from year to year.

I discovered that the gentoo also relies on pebbles to form the main weight of the nest when I saw the birds squabbling among

What sight could be more pleasing to an ornithologist than these massed penguins in their native habitat on the Jasons

Not the most orthodox way to catch a rockhopper, but this little wire crook was most successful

Rockhoppers formed the subject of this delightful Christmas card which I received from the famous bird artist Robert Gillmor

themselves over individual pebbles, after which the victor of any particular battle would carry the one stone up the beach to the chosen site for the nest, sometimes up to a mile away, a trip which was repeated countless times. It struck me that building the pyramids must have demanded the same devotion to duty. The gentoo nest is certainly just as much a masterpiece of construction as the Egyptian monuments, built with a pronounced dip in the middle for drainage. So fascinated was I by this avian architecture that when I came across one nest which was 2ft 6in wide and about 9-12in high, I dismantled it and found it to be composed of some 500 pebbles. I then put it together again, I hope to the owner's satisfaction. On my return to Birdland I remembered this dependence on pebbles, so looked around for some with which to keep our gentoos happy. I was lucky to find a company which has a licence to take pebbles from the East Devon beach of Seaton and for £110 they delivered a lorryload to Bourton.

Now that we had accomplished the major part of our task we

could sit back and wait for the *Forest*. We did not have long to relax before we spied her mast on the horizon, a most exciting sight which gave us the feeling of shipwrecked mariners at the end of a long, enforced isolation. I must have been under the spell of the island if I thought for one moment that our work was now over, for here were sixty-six penguins corralled a mile-and-a-half from the shore with their packing cases not yet assembled on a ship anchored out at sea. However, a small boat came out for me and, once on the ship, I came to an arrangement with the good Captain Jack Solis and his crew whereby I paid them a little extra to help with making the cases and fetching the birds. This worked out well and in no time at all our impromptu assembly line had put together all the boxes and we set off to fetch the birds. It was a pretty exhausting walk from the corral to the shore, stumbling back with heavily laden boxes, but we finally made it and, after making sure that we had enough fish on board to keep the penguins happy, we set sail for Stanley.

Our arrival at that port created quite a commotion with the townspeople turning out to see us as if they had never seen a penguin before. Once disembarked I was summoned to the Colonial office to pay duty on the birds in order to be able to bring them out of the Falklands. As well as penguins I was also bringing back some geese for Sir Peter Scott at the Slimbridge Wildfowl Trust. The very strait-laced official scanned the list of rates of duty payable, then announced that I owed him £1,200. 'Ah, sir,' I replied, equally straight-faced, 'I'm afraid that I've come on a different matter. You see, I've taken stock of all the birds on my islands and estimate that there is forty million pounds' worth of wildlife there which I would like to sell to you. I'm prepared to settle for half the duty owing to me on these birds and I'll take the money now if I may'. At first he was flabbergasted but then realised that I was not serious and he relaxed a little as I paid up, thus giving me clearance to take the birds off the Falklands.

Farfetched as my little joke might have been, it is amazing how many people now think that I should plunder the islands for commercial gain. I have been offered all sorts of fantastic sums, up to £50,000, to provide certain people with stocks of birds, but have only accepted commissions from those I know to be genuine conservationists who wish to preserve the species and breed from them rather than seeking to exploit the birds in any way. Often I make reciprocal arrangements with zoos throughout the world whereby I provide them with penguins say, and they give me a type of bird which hasn't previously been kept at Birdland.

Leaving Stanley for South America with my penguins

Two examples of the specially constructed boxes which I had transported out to the Jasons in order to carry my penguins home

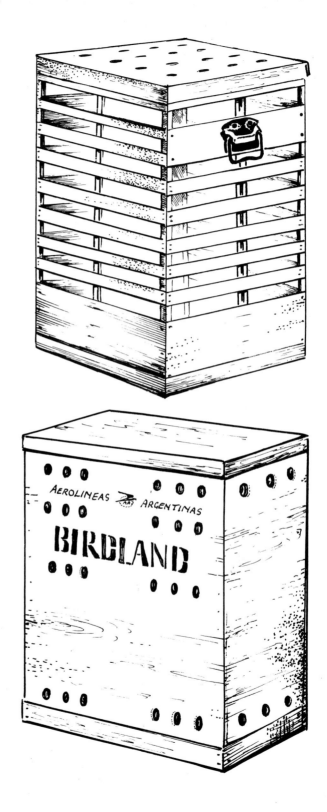

(*Opposite*) As usual the rock-hoppers are in the picture again. This time they are hiding behind a king shag on the Falkland Islands

carefully cleaned and fed, I at last felt that I was on the final lap. I had arranged to keep the birds overnight in the Montevideo zoo and presumed that all details would have been taken care of. On arrival, however, I was surprised to learn that a new law had been passed in the Argentine, my next stopping place, to the effect that all livestock landing there should give three days' notice of inspection before a veterinary certificate could be issued to clear their passage. It seemed best to stay put, give notice of my impending entry to the Argentine authorities and book another flight in a week's time, but I was now left with the problem of keeping my sixty-six penguins in a temperature of 103°F (39.5°C) for seven more days.

I went straightaway to the zoo to try to explain my plight but drew a complete blank at first as no one there appeared to speak English. Just as I was beginning to get worried the daughter of the zoo director, a trainee vet, arrived and, as she spoke a little English, she was able to tell her father of my predicament. He suggested that I put the penguins in the duck pond, a mere trickle of water awash with green slime and inhabited by river pigs. I refused this offer as it seemed that the birds would come to less harm in their boxes. Scouting around, almost in despair at the thought of even now losing the penguins, I came across a cellar under the hippo house which was very dirty and full of old cages, but it was large and, above all, cool, which would at least give the penguins a chance of survival. I obtained permission to use this room and spent until midnight cleaning it out before releasing the birds.

Then began a daily cleaning and feeding routine which occupied most of my time in Montevideo as each bird had to be caught and fed by hand so that I received numerous bites and lacerations. It was lucky in a way that I didn't have much spare time and wasn't able to spend any money. Not that I had any to spend should the opportunity have arisen: the banks in that city only open three days a month, none of which coincided with my stay so I had no currency at all. Fortunately my hotel's owner was about to make a trip to England in the near future so I gave him English money in exchange for the meals and accommodation.

The third day in Montevideo was almost the final straw. I arrived at the zoo early in the morning as usual to find no one else there, and it soon became obvious that no one was going to come. The keepers had gone on strike—they were earning barely £3 a week—and refused to go back to work until promised more money. All the other services stopped too so that for two whole days no animal in that zoo was cleaned out or received any food—apart that is from my penguins. As soon as I realised that no fish was available I decided that I'd have to go

and get some myself. So, armed with as many secondhand plastic bags as I could get hold of, at the extortionate price of 50p each, and a ball of twine which cost 70p, I hired a taxi and told the driver to take me to the fish market on the river Plate. There I bought every bit of fish which I could lay my hands on and triumphantly went back to the zoo where I became a fishmonger, cutting up the larger pieces of fish for the gentoos and feeding the tail ends to the little rockhoppers. This ensured

My own private postage stamp, issued to commemorate my purchase of the islands, was printed by Harrisons, the company which produces all British stamps

the penguins' survival until the zoo began normal working again. I've never seen anything so remarkable as when the food lorry came back for the first time after the strike: the seals, in particular, were so hungry that they jumped right out of their enclosures and followed the progress of the vehicle round the zoo.

On the last day before I was due to set out for Buenos Aires I gave the penguins a final clean and labelled their cases ready for the short flight over the river Plate. Late that evening it was as if some sixth sense told me to look in on the zoo to check if everything was all right. Of course, it wasn't: the hippo keeper had turned on the valve connecting the hippo waste with the cellar and all the filthy water had run down through the cases of penguins. I stripped to my underpants, washed the birds yet again, swilled down the room and, fearing that the 'error' might have been committed out of spite, I stayed with the birds throughout the rest of the night.

Apart from an incident with a tough steak in the Argentine, the supposed home of beef, and a delay for three hours before our plane could take off, the rest of the journey passed without incident until I arrived at London airport where, for some reason, the customs officials refused to believe that these were my own birds and a long argument began over the amount of duty payable. Fortunately, I had taken the precaution of carrying the deeds with me and the issue was settled amicably in the end for the price of two of my special Jason Island stamps which had been printed to commemorate my buying the islands.

At last we reached Birdland and I don't know who was the most ruffled from the experience, me or the penguins. I rather think that I was as the birds soon settled happily in their new home and seemed little the worse for the journey. Yet I had at last been to my islands and determined to make another trip as soon as possible, only the next time I swore that I would travel with a bit more style.

Paradise Regained

Quite by accident I was as good as my word in that my next visit to the islands was a well-documented affair when I went with the BBC TV camera crew who were filming me as the subject of a programme in 'The World About Us' series called, like this book, 'Penguin Millionaire'. In preparation for the trip we had assembled half a ton of camera equipment, plus stores and cases, with such pre-organisation that I felt sure that little could go wrong this time.

Yet, shortly before we were due to depart, my island demon struck another blow as I received a message that the *Darwin*, the ship which was to have taken us from South America to the Falklands, had been taken out of service. Knowing that they were just starting twice-monthly flights from the Argentine to Stanley, I wrote immediately to that country's London embassy, enquiring about the possibility of chartering one of these craft or a ship. At first there was no reply and Ned Kelly, the programme producer, was beginning to get anxious as filming at Birdland was complete and he needed to wind up the whole film in time to schedule it in a good slot in the series. After much waiting and anxious phone calls I was finally told that, although no civil planes were available, the Argentine navy would be pleased to lay on two Albatross aircraft for me—free of cost!

Without wasting time we took the first scheduled flight only to find that at Rio Gallegos, situated near the tip of South America, the weather was so bad that we had to wait for another three days before we could fly to Stanley. As soon as the weather calmed we took off, flying steadily at about 200mph in the solid old planes which had been purchased from England twenty-five years before. When we were about an hour out from our destination a signal came over the aircraft's radio which read: would Len Hill like to avail himself of the opportunity of being taken to his island the next morning by HMS *Endurance*, where he would be off-loaded by helicopter, weather permitting; decision immediate.

Naturally the BBC team were a little put out when they had come 9,000 miles to film my landing on the islands and they hated the thought of bad weather robbing them of their prize. In order not to disappoint them I cabled the reply: await decision one hour, filming in progress, Hill. By this time we

There's nothing like a cooling-off dip on a hot summer's day

had arrived at Stanley and, after consulting the weather station at the capital which gave a good forecast for the area, I decided that we would be delighted to take advantage of HMS *Endurance*'s kind offer. Thus we were able to complete the filming and still have time for a surprise drink with the Colonial Secretary at Government House before setting sail the next day.

The most amusing part of the whole incident occurred when we arrived back at the Secretary's residence, still in our travelling clothes, and he told me: 'You know, Hill, we were so embarrassed about all the arrangements so kindly made for you by the Argentine that we thought it best to send a signal to the Admiralty in London to assess the possibilities of the *Endurance* taking you to the island. We couldn't let it be thought that the British don't know how to look after their fellow countrymen.'

And look after us they did. The voyage on the *Endurance* and the helicopter landing went according to plan and the whole trip was much easier than the first. Looking back, I was amazed to think that my birds and I had caused such diplomatic flutterings—I suppose we were simply pawns in some diplomatic game of oneupmanship. Whatever the reason, we did very well out of it and a successful film was made by both British and German television which has now been shown throughout the world. Only the other day I received a card from Bahrain in the Persian Gulf from someone who had seen it and expressed a wish to come to Birdland.

The film crews endured the hardships of island life without a murmur. On arrival we discovered that the original hut which Ian Strange and I had occupied had been blown away in a gale, so we had to set to and clean out the larger sheep-shearing hut to use as our base. At a later point on that trip, walking by the shore, we discovered some of the cans of old army rations which we had left on the previous visit and which must have been dispersed in the storm when the hut fell in. The 20-year-old corned beef and steamed pudding in the tins provided us with a hearty meal one evening.

Movement on the islands is difficult, especially through the bird colonies as the penguins will take any chance to lacerate the calves as you pass through them if you are not speedy, and my heart went out to the camera- and sound-men who stumbled through the birds while following me with their heavy equipment. These men, who were not naturalists but technicians, were quite awestruck by the sight of the massed birds on the island, a sight which many ornithologists would give almost anything to see. And, though I have visited the islands several times now, I always have the same feeling of excitement as I approach them.

If Birdland comes near to my idea of paradise in a domestic environment then the Jasons are certainly paradise in the wild for me. It is now my proud boast that I own the smallest zoo and the largest nature reserve in the world. There are about two million rockhopper penguins on the islands and about 500,000 gentoos: there is approximately the same number of Magellanic penguins (*Spheniscus magellanicus*), which can be recognised among the others at Birdland by the distinctive pattern of white and black bands which cross their throat and necks and reach down to the flanks, and another broad white band which loops from crown to cheeks, meeting on the throat. This large, shy penguin is a common summer resident in the Falklands where it is known as the 'jackass' because of its long, mournful donkey-like cry. It excavates to build its nest in sandy or peaty soil, burrowing as much as four to six feet deep down a tunnel which widens out to form the nest chamber.

There are also some 200,000 black-browed albatrosses (*Diomedea melanophris*) on the islands, and these birds make a fantastic sight when nesting *en masse*. They nest like flamingoes on the second storey up from the rockhoppers which are amongst them, possibly for protection. The nests are solid pillars

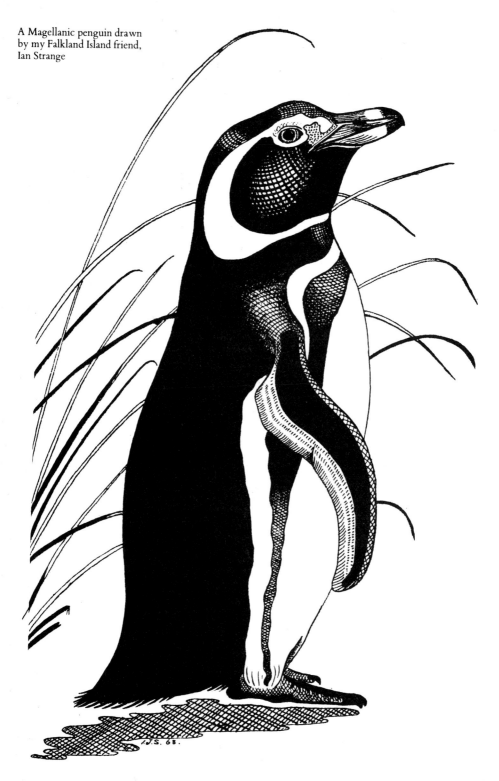

A Magellanic penguin drawn
by my Falkland Island friend,
Ian Strange

One of the most noticeable inhabitants of the Jasons, the large black-browed albatross

of mud and grass with a depression on top that holds the single, large white egg. Covered in soft grey down, with black bills, the nestlings sit waiting for the adults to bring them food, clopping their bills loudly at any stranger who comes too close.

One of the largest seabirds in the Falklands, this albatross has very long and narrow wings with a span of about eight feet. A superb flying machine, it can glide and circle for minutes without flapping its wings. The black brow from which it gained its name, a straight black line over and through the eye, gives the bird a scowling expression and its long, deep, hooked bill adds to its fearsome appearance. Indeed, the albatross betrays no sign of being put out by man's presence and we were able to walk through the colony, stroking them on their backs, without their taking any notice, apart from the frightened youngsters who vomited up a terribly oily and messy substance, the smell of which is difficult to remove from clothing.

Life on the islands is generally smelly and it is best to keep upwind of the bird colonies as much as possible. It is also difficult to hold a conversation because of the noise made by the screaming of millions of birds. A particularly loud guttural call which can be heard some distance away from the colonies is that of the king cormorant (*Phalacrocorax albiventer*), a large and handsomely marked bird of the sea cliffs. Black above with a metallic blue and green sheen, it has white underparts which extend unbroken to the base of the bill. Its colourful appearance is added to by a large orange knob between forehead and bill and blue skin around the eye. In spring it sports an untidy crest and white hair-like plumes behind the eye. The king cormorant has enormous pink feet which it uses for swimming underwater when fishing, and large flocks of this bird are often seen offshore when shoals of small fish are present.

The noise is perhaps particularly noticeable when the chicks are hatched, but noise of a different kind goes on during the mating season. One of the birds to utter strange calls at this time is the kelp goose (*Chloëphaga hybrida*) which displays a great deal in late winter and early spring. The male of this large goose is the most conspicuous bird on the rocky shores of the Falklands, having pure white plumage with yellow legs and a black bill. During display he throws back his head to show the snowy-white breast, uttering wheezy whistling notes. The female, nowhere near as striking as her partner, is mainly dark brown, barred black and white below with a white tail. She holds her head high during courtship and gives a loud, hoarse hooting call. In late October or November a grass nest is constructed in vegetation near the beach in which the female lays up to six buff eggs on which she sits, while the male unwittingly reveals the site of the nest by insisting on standing guard just a few yards away.

Eggs and goslings appear to litter the ground at all times of the year. Taking even a short walk becomes difficult with youngsters scuttling under the feet and skuas dive-bombing from on high. The great skua (*Catharacta skua*) is notoriously aggressive when breeding: both adults will swoop at intruders, calling loudly and striking with their feet as they fly overhead. Waving a stick above the head will keep most skuas at a distance but they are sufficiently scaring to make a dog run for cover. An unattractive bird with a strongly hooked, stout black bill, its calls are very harsh and guttural, reflecting an equally

(*Above left*) The creamy-white pendulous flower of the soporific moonflower, *Datura cornigera*

(*Above right*) An example of how flowers can be successfully brought from one environment to another, this *Datura cholrantha*, golden drop, was collected by me from the late Pat Delapenha of Mandeville, Jamaica, a great naturelover who planned to lay out the grounds round his house as a reserve, at my suggestion. The plan may yet come to pass in conjunction with the World Wildlife Fund but, in the meantime, I like to think that this fine flower serves as a small memorial to him

(*Below left*) The passion flower, *Passiflora quadrangularis*

(*Below right*) *Aechmea fasciata*, which comes originally from Brazil, is possibly the most popular of this attractive species

unattractive personality for, as noted in the section on the Antarctic, it preys quite mercilessly on penguin and albatross eggs and young.

When it is revealed under the mess left by the birds, the flora of the islands is quite beautiful. One moss gives a lovely scent if walked on when in bloom, rather like an eau de cologne, and I'm particularly attracted by a magenta-flowered dwarf sorrel which is found in carpets all over the island, looking from afar like a magenta-coloured cloud. Patches of a pretty, yellow, miniature calceolaria-like flower are also delightful. On my next visit to the Jasons I hope to bring back some of the more unusual flowers to see if they will grow in our climate.

I have already had some success in growing tussock grass, a plant which is associated particularly with the Falklands. Very nutritious and tasting a little like chestnuts, it is quite edible. Indeed it is said that people shipwrecked on the islands have existed on tussock grass for long periods. It is particularly nutritious to sheep and much was eaten away when the Jasons were farmed, but now that the livestock have gone the grass seems to be making a comeback.

Much of the grass was also destroyed by burning in the days when not only sheep but seals and penguins were also farmed. This inhumane practice involved slaughtering the birds and beasts, cutting them up on the spot, then extracting their blubber and oil. There are still three huge tripots on Steeple Jason which were used for boiling up the poor creatures. Weighing between 50cwt and 3 tons, the big pots were made in Scotland and transported to the Falklands in the 1860s when the land was first sold to Dean Brothers. Remaining piles of bones are evidence of the horrid work, as are long stone corridors which used to be covered with wire netting so that the birds and seals could be driven down them to the tripots and their death. I very much want to take some base huts and a mobile tractor to the islands in order to make a proper study of this awful form of exploitation and the effect it has had on the wildlife of the islands over the years. I also intend to transport two of the tripots to new homes, one to Birdland and the other to the museum at Stanley, leaving one remaining on the islands as a testimony to man's cruelty in previous times.

Burning the grass did not, I am glad to say, signal the demise of the tussock bird (*Cinclodes antarcticus*), often the first creature to greet visitors to the outer Falklands. Small, dark brown in colour with a slender bill and black legs, the tussock bird is unremarkable in appearance, but its behaviour is amazing. As soon as anyone lands on the islands the birds appear to scurry round their feet, even perching on people as any disturbance of the sand or rotted weed reveals small flies or grubs which the

(*Opposite*) Everyone's favourites —a group of king penguins out for a stroll

One of the huge tripots left
behind on the Jasons from the
cruel days of penguin and seal
farming

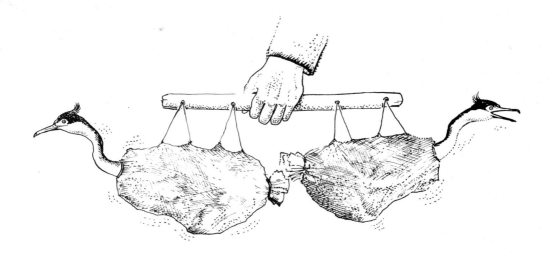

This unusual way of transporting cormorants by tying them up in sacks suspended from a stout stick was certainly effective on the Jasons

bird snaps up and eats. It is so tame and inquisitive that one day, when I was having my lunch in the hut, two came and perched, one on each toe cap of my shoes. They did not seem to want any of my meal although it is said that they are so opportunist where food is concerned that they will enter farmhouse windows to take crumbs from the table, and will take pieces of regurgitated fish spilled by the king cormorants when feeding their young.

The tussock bird is numerous on the Jasons where there are suitable beaches for food and where the ground provides enough holes for nest sites. The nests are usually found under an outcrop of rock, made of grass and lined with sheep's wool and fine grass. Very pugnacious towards others of their kind, the tussock birds can often be seen chasing each other in flight or on the ground, especially in the breeding season, and continually uttering their sharp, squeaky calls.

A final Falkland Islands bird worthy of mention is the flightless steamer duck (*Tachyeres brachypterus*) which is unusual as it does not occur naturally in other parts of South America. It is known locally as the 'loggerhead' or 'logger' because of its heavily built head and beak. It is a large bird with ridiculously small wings for its body and is incapable of normal flight. The name steamer duck refers to the bird's habit of 'steaming' rapidly over the water, its fast wing beats and energetic paddling producing a tremendous spray, while its breast ploughs the water like a boat—a most unusual sight. Each pair of 'loggers' fiercely defends their own stretch of coastline against other couples. They have no natural enemies apart from sea lions and are therefore the dominant birds in the dense beds

My first visit to the Jasons with naturalist Ian Strange; note the surrounding clumps of tussock grass

of giant kelp that encircle the Falkland coasts. The male's loud, challenging 'chēerŏo' calls and the female's weird guttural, creaking notes may be heard at any time of day or night.

So many birds to study and so much more to discover. I haven't been able to manage a trip to the islands over the last winter, but I certainly intend to go this year if business commitments permit. My eventual aim is to establish base huts and a research team there to catalogue the birds of the area and their habits. To this end I have set up a special islands fund and hope soon to get my plans under way with the ultimate idea of leaving the Jasons to the World Wildlife Fund.

(*Opposite*) Infant's eye-view of a king penguin

Birdland Stars

Much as I appreciate the chance to study birds in the wild on my islands, I'm always pleased to come home and be greeted by the inhabitants of Birdland. I suppose that much of this pleasure, like returning to well-loved family and friends, is that I know the birds and they know me. So many of our birds are real characters, with definite, individual personalities that never fail to delight our visitors, and I'm sure that we owe much of the garden's success to these amusing and ever-entertaining members of the troupe. Indeed, I often think that in some ways Birdland can be likened to a stage show, with the vast majority of the brightly coloured birds as a kind of constant chorus to the 'stars' such as the inquisitive macaws and parrots, which are always up to new tricks.

Perhaps the most famous parrot in Birdland is Juno who, as mentioned right at the beginning of this book, has the privilege of being allowed to sleep in the house with us. A little African grey parrot (*Psittacus erithacus*), I purchased her for £30 some twenty years ago. All too soon I discovered that she had a very spiteful nature, disliked anyone going near her, and would take hard nips at anyone who came too close.

Then, shortly after her arrival, Juno caught a chill. Her eyes were affected, she was unable to see properly and promptly lost all interest in life. Now she became the object of special attention and I took to feeding her by hand with milk chocolate, giving her that individual care to which she wasn't accustomed. Very soon she was on the mend and, as her eyesight improved, she began to recognise me as the fellow who was pulling her through. She became very attached to me and was soon flying about the garden, coming to me as soon as she heard my call.

Juno's mate, George, is called after the person who gave him to us. He is a rare example of a pet bird which has settled down well in the garden. It's now a principle with us not to accept birds from other people as they hardly ever turn out well. It's astonishing how many people do want to offload their birds on us, particularly mynah birds, African grey and Amazon parrots. Everyone likes parrots, it seems, but few know how to look after them. All too often, children see large birds in pet shops and persuade their parents to buy them, only to find that such creatures need more attention than anyone in the family can

Juno, the most famous parrot in Birdland, takes a spin watched by one of the macaws

give after the first excitement has worn off. Birds coming from such homes are often uncared for and have ill tempers or are in bad condition and would be a liability to the rest of our birds. In order to try to ensure that such birds do not suffer, however, I keep a register of people with birds to give away or sell, and people wanting particular species so that I can put the two sides in touch with each other should a happy coincidence occur. I call it my 'marriage bureau' and we've been very successful in several cases.

The marriage of George and Juno was a real love match and the two are always together. Although very different in temperament they get on well and, like human couples who have been in each other's company for many years, they agree to differ. African parrots were the favourites of Roman emperors and it's easy to see why as they are so tame and faithful. Whenever I've been away and arrive home late I always find Juno and George waiting at one particular spot so that I can pick them up and pop them back in their cages, side by side in the tropical house.

In the early days Juno once caused quite a stir on a local farm.

(*Opposite and above*) Juno is so
tame that she loves this game we
play together: up she goes and
down she comes, legs in the air,
eyes half closed with pleasure,
secure in the knowledge that she
will land in my hand

At dusk, when it was time for his hens to be secure in their house, the farmer was surprised to find that they simply refused to go in. He tried all ways to coax them to no avail, until at last he looked inside the house and saw a little grey parrot holding them at bay. She came home quietly enough when the irate farmer sent for me, and I always suspect that she was as scared as the hens were! While she is still reserved with strangers, Juno will let me do anything with her. As Johnny Morris mentions in the Foreword, she loves lying on her back in the palm of my hand and being spun round like a top, or being tossed in the air and caught like a ball.

Children love to see her antics and I use their interest in her to try to teach them a few easy points about bird recognition. I think it is important, especially for the youngsters, that they distinguish between some of the birds as they go round the garden and perhaps in this way a wish to learn more about the birds will be fostered. One of the simple facts which I explain to most visitors, grown-ups included, is how to tell the difference between parrots and macaws. Although these birds are members of the same family and look very similar, it is easy to know which is which if you remember that parrots have short tails, and macaws have long ones. Cockatoos, of course, have crests. Then there are related members of the *Psittacidae* family such as the parakeets and lorikeets, but these need more advanced study than can be gained in a visit to the garden, or indeed in this story.

The free-flying macaws possibly attract the most attention in the garden, partly by their gaudy plumage and partly because of their distinctive personalities. Charlie and Ann, the blue and gold macaws were, as described earlier, the first of the garden birds to be set at liberty. Originating from the Amazon, these striking birds are extremely gregarious and their piercing yells can be heard all over Birdland.

Charlie and Ann live in an acacia tree, a place seldom visited by the hyacinth macaws, Leah and Mac, who seem to prefer a somewhat grandiose style of Gothic architecture for their home, as shown in the colour photograph opposite page 35. The hyacinths are reputed to be the cleverest and most sensible birds of this family, which is certainly the case with these two. Conceited and exhibitionist, Leah and Mac are also very loyal, both to each other and to me.

I obtained the pair from a zoo which wished to dispose of them as Leah had bitten a keeper's finger so hard that she had broken a bone. This was a little off-putting but I believed that she would come round with a little affection and, when I went to have a look at the birds, I was introduced to Mac, the more docile of the two, so I decided there and then to purchase them.

Unfortunately, I was quite unable to begin to try to gain the birds' confidence as Leah developed pneumonia as soon as she arrived at Birdland. She had been confined to very dark quarters and, despite the fact that we put her through the usual acclimatisation process, she became very ill, possibly through shock.

Leah refused to eat and seemed determined simply to fade away. Upsetting as this was for us, it was even more unsettling for Mac, as marriages between hyacinth macaws last for years, often a lifetime, and he too became disgruntled. Apart from putting Leah in a warm place I was at loss as to how I might save her. Eventually I asked the advice of my old friend Reg Partridge, the man who has given me more help than anyone else in the bird world. He recommended that I feed her on ash bark, apparently a most nutritious food. It has been proved that,

Here comes Mac on the prowl, looking for mischief

in extremely cold spells, when rabbits cannot obtain any other food from the frozen earth, they exist for long periods on ash bark alone. I fed this to Leah and, amazingly, in three months she was completely recovered. I was almost as grateful as Mac for this and the three of us have since remained devoted friends.

These two recognise my footsteps even from the other side of the garden and let out their raucous call of recognition. Then they come to find me, tip-toeing along like two old ladies on an outing. As soon as we spy each other we stand and stare to see who will make the first move. 'Hello there, hello there,' says Mac with his head on one side, then starts waddling along the ground towards me at a great rate. I extend my foot slightly and he steps on the toe, climbs straight up my leg and on to my shoulder, saying 'Ah, ah' and delighting in the attention.

Leah and Mac often accompany me on gate duty for an hour or so until they become bored. As anyone who has seen our 'Birdland Story' film will know, these two are always up to mischief, frequently going 'out of bounds' as they have such curious natures. They often raid the other birds' food and are not averse to nipping a button from a visitor's coat if it attracts them. Extremely intelligent, they have supremely heightened senses, acting as a built-in early warning system for the rest of the garden's inhabitants. If a predator, such as a buzzard, or even a lone seagull flies over, just a speck in the sky hardly distinguishable to the human eye, then one of the macaws will spot it immediately and let out its deafening alarm call.

Photographs of Leah and Mac have now gone round the world. Franz Lazi was first introduced to Birdland when searching for colourful subjects to photograph for a trade calendar and is now almost as much at home in the garden as I am. He has shot many thousands of feet of film with me, here and on foreign trips, and is looked on as one of the family. His outstanding colour pictures in this book have been used by us on postcards, greetings cards and even on special wrapping paper which we use in the Birdland gift shop. And I understand that the German company which first commissioned Franz's bird studies has now used the prints on wallpapers and silk headscarves.

The most photographed Birdland member from the point of press and publicity is undoubtedly Frederick, the pelican, as he is now the mascot of the Central Flying School at nearby RAF Little Rissington. Flight-Lieutenant Frederick is quite a celebrity, always attending dining-in nights at the officers' mess and being introduced to special guests such as HM Queen Elizabeth The Queen Mother, who is Commandant in Chief of the Central Flying School, Rissington (see page 133). At times like this, Frederick receives his typed instructions like any other

(*Opposite*) One of our friendly macaws ready to shake hands with any visitors

serving member of the staff and next year he will eligible for the status of squadron leader.

I shall never forget the first night that Frederick went on parade. It is a custom at Little Rissington for the courses moving out after instruction to have a farewell party and to be presented with a little gift. As the pelican is the emblem of the squadron, someone had the bright idea of giving one to them, so they rang me at three o'clock on the afternoon of the dinner to ask if I had a pelican for sale. Of course, I wasn't prepared to sell Frederick, but when they explained what they wanted him for, I was quite in agreement for him to become their mascot. So we washed his feet and generally spruced him up, then Rosemary and I arrived at the officers' mess half an hour before the dinner began.

The essence of the plan was its surprise element so we were ushered up the back stairs to a balcony overlooking the dining room, and there we had to sit throughout that long dinner, doing our best to keep Frederick quiet and happy. At one stage we thought that the speeches would never end and Frederick was so excited at such strange goings on that he couldn't help answering calls of nature, so that by the time the affair ended Rosemary and I were beginning to smell like pelicans ourselves. Luckily, the moment when he was presented made up for all our discomfort. The screens were taken away, the balcony was floodlit and Frederick stood up at the front, bowing to left and right as he looked down at the two hundred or so assembled officers. Of course, his presence was a masterly stroke and the applause rang out for several minutes. We then took him down to be introduced to everyone and ever since he has been invited to all ceremonial occasions.

One event which was a little more difficult to organise was when it was necessary for Frederick to be introduced to his new CO at dinner. There could be no surprise entrances from on high this time but an official presentation in which our Flight-Lieutenant would have to walk out to meet his chief. So we devised a plan, unknown to the officers, by which Fred would be kept a little short of food during the day and I arranged for one of the waiters to place a bowl of the pelican's favourite sprats under the top table. We put Frederick in a special box at the door of the kitchens and, when the moment came for him to be presented, out he stepped, standing five feet high, a memorable sight, and began to walk straight between the tables towards the CO. He, of course, could smell what no one else could—that bowl of sprats under the man's feet. He stopped directly in front of his new chief, then slowly bowed his head, a move which virtually brought the house down as the diners had no idea that he was really bending his head to get at the sprats!

Frederick is always on parade at the 'At Home' day at

Cheltenham for which he has his own special conveyance, a
military police Air Force Land-Rover with a light plastic floor.
He behaves wonderfully when I'm with him, but tends to
become excited and snap out when the airmen are clicking their
heels and saluting. It's a fine sight to see him riding along with
the rosy pink sack under his huge beak fully extended like a
parachute. Pelicans in flight do look exactly like a bomber
squadron, wave upon wave sweeping across the sky in V
formations.

So clumsy on land, pelicans in flight are a marvellous sight,
especially when many birds are massed together. Frederick does
not have his wings clipped and is quite capable of flying but we
try to make sure that he's always kept in a fairly confined space
without enough room to take off. He's always ready to try
though and he will start his noisy, preparatory take-off
movements at any time, drawing together his large feet and
flapping his huge wings in an effort to catch an air current and
get off the ground.

Of course, there are always some people who like the unexpected to happen, especially newspaper reporters out for an unusual angle on their story, and on one occasion I was asked if Frederick could be photographed out of his little corral. It was a particularly windy day and I could see that the pelican would be off, given half a chance, so I asked a line of pilots to stand in front of him to prevent his escaping. He's quite a fearsome sight when 'revving up' for take off and I suppose that his actions simply scared these flying aces for, as soon as he began belting down his runway, they immediately scattered and off he went. I was convinced that we would never catch him but fortunately he flew towards a wall and hesitated, almost as if he believed that there was wire or something on top to stop him, and we were able to catch up with him. Never again shall I pander to newsmen or trust daring pilots!

We heard recently that RAF Little Rissington is to close and the Central Flying School will move to the base at Cranwell in Lincolnshire. Frederick will remain the mascot, and somehow or other we hope to be able to transport him to Cranwell for ceremonial occasions.

We've had lots of fun with some of the other large birds which have taken it into their heads to have a little trip outside the garden. In the early days a couple of cranes were always out and about, so we came up with a method of catching them by 'going fishing'. We would go off with a fishing rod but, instead of a hook at the end of the line, there was a little lead weight, and we became quite adept at casting for runaways with this. The weight would fall on top of the bird, just in front of the wing and prevent flight without causing any harm. Accurate casting is nowhere near as easy as it sounds and we had lots of fun perfecting the technique which we also used on escaped flamingoes. These birds, however, often ended up on lakes so we got wet several times when bringing them back, but I'm pleased to say that this rash of escapes has virtually ceased.

Our most spectacular escapee was one of Birdland's most unusual inmates, the great Indian hornbill (*Buceros bicornis*). Found up to an altitude of 5,000 feet in the Himalayan foothills, this fearsome-looking bird is five feet long with the same amount of wingspan. It is unique in that it is the only remaining bird with eyelashes, and its main distinguishing feature is the huge bill, topped by a large, brightly coloured casque. The way to tell the difference between the two sexes is that the male has a red iris while the female's is grey. It is worth noting that the irises of all female birds' eyes are lighter than those of the male, and this is particularly easy to spot in the large hornbills.

Birdland's hornbills are the only pair ever to have been kept in captivity for any length of time. They are about eighteen

(*Above left*) We are most proud of this banana tree, seen here bearing fruit

(*Above right*) The exotic-looking flower of the *Monstera* whose slashed leaves are equally attractive

(*Below left*) *Ananas bracteatus striatus* is the botanical name of the variegated pineapple which grows profusely in the tropical house and is now becoming increasingly available in this country from leading nurserymen

(*Below right*) The scarlet flowers of the lovely *Clerodendron speciosissiumum* do particularly well in the moist conditions of the tropical house but are equally at home in warm greenhouses and can be raised from seed

years old now and may possibly live to about forty. Perhaps the most curious feature of their life-style is when nesting, when the incubating female is imprisoned by her mate in the chosen site. The egg is laid in the cavity of a tall tree and the male literally walls her in by building defences round her made of droppings, regurgitated matter and mud, leaving a large hole through which the female's bill protrudes. This defence is put up in the jungle as a protection against predators, mainly monkeys and it is almost impregnable. The female hornbill is fed by her partner throughout the long period of incubation, as many as forty feedings an hour having been recorded. When she emerges after two to four months' confinement she is a sorry sight, just beginning to replace her feathers as all the original ones have been used for nest lining. We thought that we had achieved a real first when our hornbills, which have been nesting in this manner for some years, actually produced a chick. Unfortunately they were so new to the experience that they did not know what to do with it, or so it appears, as the male broke into the nest and killed the chick, either through jealousy or by treading on it inadvertently.

It was this same male giant hornbill which made such a startling escape. One year we were having some work done in the garden close to the hornbills' aviary. Someone carelessly placed a plank of wood under the metal support of the aviary which holds the mesh in place on the top, and another person then leant on the plank which fell down, prising the metal as it went, so that the staples flew out of the mesh, a corner of light was exposed and, equally quickly, out flew the huge bird. Hornbills do two or three swishes of their wings per hundred yards, making a noise quite as loud as a swan's wingbeats, and up he flew right to the top of the acacia tree. We immediately secured the top of the aviary to keep the female in on the assumption that the male would not want to be too far away from her, but her presence, or rather lack of it, did not seem to bother him as he flew off about two miles away.

Our main fear was that anyone seeing such a fearsome-looking creature in their back garden would panic and shoot him. We informed everyone we could and kept a constant check on his movements via phone calls and personal observation. Stow Fair was on, some four miles from Bourton, and he must have been attracted by the bright lights and music as he went for a night out there. It was hopeless to try to catch him amidst all the amusements and crowds so we had to content ourselves with knowing that he was safe for the time being and keeping a track on his movements. We heard of his position one day from a camper who felt that he had had a nasty experience. Waking up one morning to a soft English day, he heard an

(*Opposite*) See how the brilliant coloured plumage of these macaws is shown off to advantage as they pose on a ladder

Imagine the camper's surprise
when he saw the great Indian
hornbill

amazing sound coming from outside his tent and, drawing back the flap, came face to face with what he took to be some prehistoric creature—our hornbill.

The bird was out for ten days without a chance of being caught but, by this time, we knew that he would be hungry. We made a trap cage some six feet square with a big, loose top, and put both the female and a tin of cherries inside. It was autumn, when the blackberries were ripe, and we soon received news that he had been sighted feeding off them in a nearby field, so off we went with all our paraphernalia, including the hen bird. We found him in a large field of some forty acres with a coppice at each corner, and set up the trap cage at the opposite end from where he was having his supper. Backed by the setting sun he looked a magnificent sight and I imagined how he would look in his jungle home. I held the female's head up a little and, sensing his presence, she let out a long, searching call. Immediately alerted he replied, and came swishing over the field at such speed that we hardly had time to dive down behind the hedge for concealment. He alighted on top of the wobbling trap cage which keeled over with his weight and we feared that the top would go down before he entered. However, his caution did not last very long for, as soon as he spied the cherries, down he went, the lid closed and we had him safe and sound.

What an episode that was. I'm glad that we don't have such excitements every day, but such an incident does serve to illustrate that life at Birdland is never dull.

A Day in the Life

People sometimes ask me if I ever get bored running Birdland, day in day out, every day of the year apart from Christmas Day when we close to the public, although we still have to feed the birds then and make sure that everything is all right. How can I be, I ask in return, when I am doing what I have always dreamed of doing, when my work is my life and my life my work. I gave up my position in the construction business on my son Richard's wedding day and he now runs it, although I still take an active interest. This is a decision which I have never regretted as it allows me to devote all my time to the birds. I'm lucky, of course, in having a splendid family and staff who simply get on with the job of running Birdland when I'm not there, which allows me to visit the Jasons and make other trips, but on the whole I'm happiest when I'm back in the garden, with countless demands on my time, working twice as hard now as I did when a much younger man some twenty years ago.

On a typical day at Birdland I'm usually up at about 5am in the summer and take my first walk around the garden before anyone else is stirring to make sure that everything is in order, particularly to note whether any of the birds seem a bit off colour. The soft-bills especially have to be watched closely to ensure that they are not ailing, as they are not so resilient as the hard-bills and need softer food given to them in large quantities more frequently. One of the signs I look out for is whether a bird is resting on one leg or two. It's an odd fact but a bird in good health will nearly always stand on one leg and if its weight is evenly distributed on both legs there is probably something wrong. The next tell-tale sign of illness is a dirty mandible and then the wings begin to droop a little with the primary feathers just showing. At this stage it is almost too late, so it's essential to have spotted the problem before then.

If a bird is seen to be under the weather we take it in to the warmth straightaway. Like humans, birds appreciate heat. We have a special hospital area at Birdland with one room where we can build the temperature up to 100°F (37.7°C). This room is divided into cubicles, each one containing a perch with a light shining constantly on one end of it. It's interesting that whilst the bird is ill it stays under the light but, as soon as it begins to feel better, it gradually moves away from the heat. At that stage the temperature is lowered slowly by degrees.

Cocky, our famous cockatoo, looks as if he's seen something interesting in this tree trunk

It's this constant fight against disease which occupies a lot of my assistant, John Midwinter's, time. An ex-carpenter like myself, John has had no scientific training in looking after birds, just what he has picked up through working with them over the years. But he does not need any official recognition as he is one of those rare people who understand birds instinctively and know exactly how best to treat them. Living on the premises as he does, John knows all the birds intimately and can soon tell if anything is wrong. We have very few failures in our fight to keep the birds healthy which springs, I feel sure, from our attention to detail and personal knowledge of the birds. We sometimes have to resort to unorthodox methods but these usually work well, mainly because the birds have confidence in us and sense that we are doing all we can to help them.

A good example of this kind of master-pupil relationship leading to mutual regard was when Cocky, the friendly sulphur-crested cockatoo, broke his leg in an incident unobserved by us and then fell in a decline. I took him to the vet to be told that the leg would never mend and would have to come off, a last resort which I couldn't bear to think of. I took Cocky back home and, instead of trying to set the broken leg in some way, I simply put him on a sloping perch which we fixed up against a wall. The perch had a large knot at the right height on which the bird could rest his good leg and not slip, leaning his weight on the perch for support. During the weeks he was unwell, I personally fed Cocky six times a day, talking to him and massaging the toes of the foot which bore all the weight as well as the damaged one. I'm sure that the bird knew that I was willing the leg to mend and, within fourteen days, the care paid off; with no recourse to splints the broken leg healed completely, the only sign that it has ever been broken being a slight swelling in the joint.

Originally from New Guinea, Cocky has now reached the grand old age of thirty. A bird without any vices, he is always popular with visitors and the villagers. Pure white, with a large, erect yellow crest, his familiar form can often be seen making a call on local tradesmen. The greengrocer is perhaps a little too kind as he doesn't seem to mind Cocky taking a bunch of grapes, a habit which I try to discourage! The cockatoo loves playing with children too and is very fond of a neighbour's small boys. One day, thinking that it was a new game, he flew over to this family's garden and removed all the pegs from the clothes on the washing line. I'm afraid that this caused quite a fuss, but the bird can hardly be blamed as it was made welcome and fed when its presence was a novelty.

An even more unpleasant incident occurred with one of the macaws. Always friendly, this particular bird was often found playing with the children of nearby families. One day he disappeared from the garden and when he hadn't been sighted for three days we had almost given him up for lost. Then, the next day, an angry parent arrived to tell me that we were harbouring dangerous birds as one had bitten her little boy and that she was going to report us to the police. We take all such charges seriously but few have any foundation. On this occasion, the macaw had been playing with her children in the garden and had followed them home. Thinking that they could keep him as a pet, the family hadn't bothered to let us know, and it was only when he gave one of the children a peck in boisterous play that anyone objected!

I'm pleased to say that this sort of incident does not happen very often, especially if balanced by the vast numbers of people visiting the garden every year. Some intrepid souls seem to take delight in enticing the penguins out to go for a walk by the river which runs past our gate. This makes a nice picture and

Schoolchildren love to come to Birdland as they can really get to know the birds instead of just viewing them in cages. Here one of the macaws says hello to a young visitor

doesn't do any real harm although I try to discourage it for obvious reasons. Others, less scrupulous, have been known to try to walk off with birds, but they have always been discovered in time. Someone once came running to me at the entrance to Birdland to tell me that a parrot was being wheeled away in a pram. Sure enough, when I caught up with the culprits, the bird was perched under the hood with a raincoat forming a tent to hide it! But it is my proud boast that we never lose anything from the garden—even the visitors come back.

How can I be bored for one moment with such events happening all the time? I must admit though that most days follow a more mundane routine and that, after my first round of the garden, I'm back in the house by 6.45am to make a cup of tea for my wife, and together we listen to the weather forecast on the radio, as the need for fine weather obviously plays an important part in our daily life. At half past seven I go out again to let the staff in to start the food preparation and general cleaning which is a most important—if unglamorous—aspect of our work.

Even at this time in the morning I've had one or two shocks. Going to unlock the gates one day I saw to my surprise that there were already about 150 people gathered outside, although

130

To
Mr Hill
a big
Thankyou

Pictures about our
visit to Birdland.
From the Infants (age 5,6 + 7)
at St Mark's Primary School, Cold Ash

Birdland does not open till 10am. I noticed that the crowd appeared to have an unmistakably Continental air, but couldn't understand why they were there at such an unearthly hour. One of the party started waving excitedly and, as he obviously knew me, I went to see what it was all about. It turned out that they were members of a Swiss aviculturalist society who had been present at one of my film shows in Bavaria and had been so impressed by what they saw that they had decided to come over and see Birdland at first hand. They had chartered a plane from Zurich, had just arrived, were staying over one night then going back! I was delighted to be able to entertain them briefly before they left, but only wish that they had given me some advance notice, as similar incidents are occurring all the time and we have great difficulty in fitting everyone in.

I receive over 500 applications a year to show films at various ornithological societies and schools and I try to limit this to two performances a week, mainly in the evenings, but often this tips over so that I'm out sometimes four nights out of five. The films have been shown all over Britain and in most western European countries. I've also taken them to the major South African cities, as well as to the United States, South America, the West Indies and the bases in Antarctica.

It may not be a Picasso, but I like it! This drawing, like the one of the penguin which is shown on page 111, is from a special book which the children of St Mark's Primary School, Cold Ash, compiled for me after coming to Birdland

(Opposite) HM Queen Elizabeth the Queen Mother has a word with Flight Lieutenant Frederick, the mascot of the Central Flying School

(Over, left) A scene and characters well known to all who visit Birdland. Here I am seated on the lawn in front of Chardwar with Juno on my arm, Mac on my knee and Cocky on the seat, with the flamingoes in the background

(Over, above) As evening falls, the last few visitors take a look at that part of the garden which houses the macaws. The lecture room is in the background

(Over, below) And a more peaceful view of Chardwar, my Elizabethan manor, after all the visitors have gone and the birds, apart from this lone flamingo, have gone to rest

It's easy to see the response generated by these film shows when I open my daily post at breakfast. Our mail bag is colossal at times, often fifty to seventy letters a day, some with hardly legible addresses from many different, faraway countries. One in particular holds pride of place in my scrapbook. Sent from Paraguay, it was simply addressed to 'Len Hill, Birdland, England', and the Post Office delivered it without any delay!

During our particularly busy period, from Easter to September, breakfeast time is the only half hour of the day I am able to talk to my wife, Nell, in peace and quiet until the garden closes at about 7pm. She is responsible for running the successful gift shop which we have on the premises, to which people flock at the end of their visit to take home a small souvenir of Birdland, ranging from a picture postcard of Mac and Leah to a £36 book on parrots which sells very well. I leave all the ordering of stock to Nell who very much enjoys dealing with this side of the business now that our family have left home. Looking at the price of some of the exquisite pottery birds and figures which are on sale in the shop, we often pause to chuckle at memories of the days when my total wages were fourteen shillings a week and we could hardly afford the necessities of life, though we are still as happy now as we were in those early days of our marriage.

By 9 o'clock I am in the office, checking up on the day's appointments with my secretary and answering any urgent letters. Meanwhile the staff are at work in the garden, cleaning up all the rubbish and sweeping the paths clear so that all is ship-shape for the public when they start coming in an hour later. I jokingly say that at 10 o'clock we let the real animals in, but by and large our thousands of visitors are extraordinarily well behaved and there is very little vandalism in the garden. The average daily attendance is over 1,000 but on one never-to-be forgotten day we coped with 11,000!

I think that people are thoughtful because we encourage them to be. Obviously in a garden of this kind with so many precious charges we have to have some rules, although I hate officialdom and try to make sure that people are told what they can and cannot do as nicely as possible. So, instead of authoritarian notices saying 'No smoking' or 'Please do not walk on the grass', I try to get the message across in a lighter vein with an illustrated placard by the carp pond which reads 'These fish do not smoke' and another, on the lawns, saying 'Your feet are killing me'. Corny maybe, but such notices seem to work like a charm and children in particular, who would probably be some of the main culprits, store them up for repetition when they return home or to school.

I treasure one conversation in particular which I had with a

bright little girl who came up to me as I was planting out pansies by the side of the pool. Gazing solemnly at the notice proclaiming that 'These fish do not smoke' she said rather earnestly 'Please sir, I didn't know fish did.' Of course, I had to reply in a lighter vein that some fish are also smoked which set us off on a lengthy discussion of these items.

I always look upon children in the garden as some of the most rewarding visitors, partly because they are keen to learn and partly because they are so appreciative. I try to make sure that I make my comments on the birds sufficiently interesting and entertaining so that they learn as much about life's pleasures as ornithology. This approach certainly seems to pay off as we have a constant stream of school parties during the summer months, and I'm delighted when so many of the children take the trouble to write to me, telling me how much they have enjoyed their day at Birdland.

One amusing incident occurred with a young girl who had obviously visited the garden with her school and now wanted to show it to her parents. Up to the gates she pranced, still wearing her blazer and straw boater, all very important as she was in charge. Behind came two very properly attired people, the father still sporting his city suit and bowler hat. Outside the Birdland entrance gates are notices welcoming visitors in many different languages as I believe that even such small gestures help cement international relations. There is not enough of this type of thing done in this country as far as I am concerned. The oh-so-correct father spotted a sign in Russian and, in a very bored voice, asked why it was necessary to go to such lengths. 'Don't be so silly, Daddy,' piped up the little girl, 'these birds come from all over the world and understand many different languages, so there have to be notices which they can all read!'

Of course, it is not only children who bring me unexpected enjoyment. Only a few weeks ago I watched as an elderly school mistress chaperoned her party of girls round Birdland. Once inside the entrance she began by informing them that 'this garden is run by an eccentric old gentleman', then going on to say 'Isn't it nice of him to spend his money for our enjoyment?' I had quite a long chat with her later but didn't reveal that I was the 'old gentleman' or that the garden gives me just as much pleasure as it gives all the visitors.

I see a fair amount too of the other side of human nature and always try to turn any snobbery, flattery or downright rudeness to good account. Once, when the half-crown was still currency, I was wearing an old smock while cleaning the windows, when a man tapped on the glass and asked if I could point out the owner of this establishment to him. Certainly, I agreed, and at that he tipped me half a crown. Chuckling to myself I put the

(*Opposite*) Charlie and Ann, the blue and gold liberty macaws, are ready to turn in for the night, pulling up the grill on their barrel home for it to be clicked into position

SEÑOR LEONARD HILL.-

BIRDLAND.-

ENGLAND.

'Good shooting by the Post Office' is how I label this envelope in my scrapbook. The letter from South America, with such a scant address, took only a matter of days to reach us

money in the World Wildlife Fund wishing well and went on my way without giving him another thought.

Thankfully, by contrast, the great majority of people, whatever their rank, are far more direct in their dealings. I am called upon to address lots of societies and advise on many projects connected with birds, and invariably I gain as much pleasure from the meetings as do my audiences. I shall never forget the time when I was summoned to lunch with one of the foremost ladies in the land who was interested in my activities. I received an invitation from her secretary stating that lunch would be at 1.05pm, which made me stop and think a little at the preciseness of the timing. I motored up to her estate on the day, arriving in front of the large iron gates at 12.55, so had a few minutes' break before tootling up the drive at exactly three minutes past one. The secretary, feeding some birds by the steps, immediately looked at her watch, exclaiming, 'Just on time, how marvellous.' After greeting her I went to the car and produced a bunch of orchids which I had picked that morning for her ladyship, and I wanted to make sure that it was all right to present them. 'Orchids, how stupendous' she replied and took me to meet her mistress.

Feeling a bit of a fool clutching the flowers, I clasped my hands behind my back as I followed her upstairs and down long

corridors. Around us dogs gambolled, almost knocking us over. Then, the drawing room door opened and I was introduced to my hostess. Bringing my hands from behind my back, I ventured, 'I've brought you a small nosegay, Ma'am', but all that remained of my orchids was the paper and a few bits of fern: all the flowers had been knocked out by the dogs! 'Not to worry,' shouted the secretary, and off she went, clitter, clatter, down the hall to retrieve the blooms. When she returned with them, seeing my distress, the lady asked for 'the second presentation please', which I obligingly made. As she turned to put the orchids in water she asked if I knew their name. 'Oh yes,' I replied, on safe ground at last, 'It's a very old variety called President Wilson.' 'What was that you said?' she asked, turning swiftly back towards me again. '*President* Wilson Ma'am,' I muttered, wondering what was wrong now. 'Ah' she said, heaving an apparent sigh of relief, 'Thank God it wasn't Harold!'

After making sure that everything in the garden is running smoothly, I usually take a walk down the road to see how things are in our gallery. For many years I have collected bird and animal paintings building up interesting examples of the works of such well-known wildlife artists as Thorburn, and it seemed a shame that these were not on view to a wider audience. I also

Here I am pointing out an item of interest to a party of children. I try to make their visits as rewarding as possible, teaching them something about human life as well as wildlife

'Dear Mr Hill I Hope I will Be seeing
Yow again I likeid The film ot
Yow showed Us best ond all of
ohe things. Love from Giles'

thought it would be a good idea to encourage the works of
living artists who found if difficult to get their own exhibitions.
So, when an opportunity came along to acquire some property
directly next to the garden, I jumped at it and transformed the
house into an art gallery. It has now been open several years and
has proved very successful. Hailed by John Tinker in the *New
Scientist* as 'At last, a permanent exhibition of British wildlife
art' housed in 'arguably the finest public aviary in Britain', the
gallery has gone from strength to strength.

I count among one of my most successful finds in this field the
work of Paul Nicholas, a young, crippled boy whose delightful
line illustrations are used in this book. Our meeting was purely
by chance, although who am I to say that fate did not lend a
hand! One very cold and windy day I came across a lady
standing in the garden, almost blue with cold and, feeling sorry
for her, invited her in for a warming drink. She explained that
her husband and son, a spastic, were still in the garden and
would be for hours as the boy liked drawing the birds. Later,
she introduced me to them and I told Paul that he could come
into the garden whenever he wished, for free. He was so
grateful for this that he offered to do a painting for me and,
rather dubious about the result, I told Paul's father of my chosen
subject as, sadly, Paul is deaf too and can only lip-read from his
parents.

I forgot all about this incident until, a few weeks later, I
received a parcel containing a letter from Paul, written in a
perfect copper-plate hand and, inside the wrapping, a most
exquisite painting which now hangs in pride of place in my
sitting room. I was so impressed with this that I called on Sir

The macaws prepare to go to bed in their barrel home

Peter Scott to ask for his opinion, and he confirmed my belief that the picture was indeed the work of a talented wildlife artist. Expressing my thanks I wrote and told Paul that if he could get together enough paintings we would hold an exhibition for him at Birdland, a happy event which took place some few months later. At that time we had great difficulty in persuading Paul to charge as little as £10 for the works and his exhibition sold out on the first day. Now he has had three successful one-man shows here and every time has sold all his paintings, although they can now cost as much as £100.

In describing that opening exhibition in the gallery, John Tinker wrote: 'Paul Nicholas deservedly dominated the exhibition. In spite of severe physical handicap, his style combines extreme accuracy with a feeling for the living, moving bird and an evocative sense of habitat.' Yet I think that the most important factor from Paul's point of view is not recognition by art critics but the fact that his success has enabled his parents to move and to buy a cottage at Bourton so that he can come into Birdland and sketch whenever he feels like it.

Throughout the rest of the day, unless there is anything special happening, I am always to be found around the garden, taking my turn on the gate, lending a helping hand to Tom, the gardener, another ex-building worker who has been with us from the start and now looks after all the grounds single-handed, or generally explaining about the birds to the interested visitors. Birdland closes officially at 6pm but we certainly don't mind if people are still in the gardens after that time: we simply ask them to close the exit gate on their way out and, in this way, we often have people enjoying watching the birds in the cool of the evening until about 8 o'clock.

Occasionally, in the evenings, we have meetings in our lecture hall for talks, film shows or just sociable get-togethers for our Friends of Birdland organisation and other naturalist trusts. But, whatever the occasion, John or I always make sure that the birds are put away safely for the night before any festivities begin. It's extraordinary how they know when it's time to go to bed. Birds live a much more routine life than we do and, once accustomed to a certain pattern, will never change it.

The macaws, for instance, live in converted barrels in the trees, just ordinary beer and wine barrels, exactly the right size for keeping their inhabitants snug and warm. The barrels never need cleaning out by us as birds never make a mess in their own home. In fact, although they may seem messy to people who don't understand them, birds are extremely clean creatures. They always have the sense to 'spend a penny' before going on a long flight, and will only mess their cages if they are not let out. I devised an easy method of keeping the macaws safe in their barrels which they could control as much as man. Over the entrance hole we placed a sort of miniature portcullis, made of strong wire, which the birds can lift up and down with their beaks. At night time, into their barrels they go and, almost like pulling across the bedroom curtains, they grip the wire in their strong teeth and pull it down, as if to bid goodnight (see page 141). We simply click a little latch in place to secure this door, then the birds are safe and sound until the latch is opened the next morning.

Ah, here comes Juno to remind me that it really is time to say goodnight and pull down the shutters. She'll not rest till I've given her the usual nightly chocolate titbit, so I'd better go and get it before she starts chattering away and prevents my doing any more work!

Bibliography

Bond, James. *Birds of the West Indies* (Collins, 1971)

Gillespie, T.H. *A Book of King Penguins* (Herbert Jenkins, 1932)

Gooders, John. *Birds, An Illustrated Survey of the Bird Families of the World* (Hamlyn, 1975)

Goss, Dr. *Birds of Jamaica* (19th century)

Grant, Karen and Verne. *Hummingbirds and their Flowers* (Columbia University Press, 1968)

Greenewalt, Crawford H. *Hummingbirds* (Doubleday, 1960)

Herklots, G.A.C. *Birds of Trinidad and Tobago* (Collins, 1971)

Kearton, Cherry. *The Island of Penguins* (Longman, 1930)

Rankin, Niall. *Antarctic Isle* (Collins 1951)

Stonehouse, Bernard. *Penguins* (Arthur Barker Ltd, 1968)

Strange, Ian. *The Falkland Isles* (David & Charles, 1972)

Woods, Robin W. *The Birds of the Falkland Isles* (Anthony Nelson, 1975)

Acknowledgements

My thanks are due to so many people but I can only list a few here and must ask for forgiveness if I miss anyone out. To all those who have helped in any way in preparing this book and in making sure that Birdland runs smoothly, please accept my grateful thanks. A special mention must go to:

Her Majesty the Queen Mother, who graciously permitted us to use her photograph; Sir Peter Scott for providing the Introduction and who, with his wife Philippa, has been so understanding about my venture; Johnny Morris for writing the Foreword and for allowing me to share in the pleasure of his broadcasts.

The Argentine Minister at the London Embassy and also the Argentine Navy; Dr Blanco, Minister of External Affairs in Buenos Aires, and Baroness Lida Von Shey, BBC South American Representative, Buenos Aires.

John de M. Severne, MVO, OBE, AFC, MBIM, Commandant, and all the officers of the Central Flying School at RAF Little Rissington who have put up with Flight Lieutenant Frederick's snapping mandibles, in his moments of frustration, without casualty.

Not forgetting those two grand old gentlemen, the late Clarence Elliott VMH and Reg Partridge, who gave kindly advice in their respective fields of horticulture and aviculture and in whose presence I spent many happy hours.

Also, my friends at the BBC for all the patience they have shown, especially from the Natural History Unit, Ned Kelly, Jeffery Boswall, Nicholas Crocker, Maurice Fisher, Michael Kendall, Christopher Parsons, Mick Rhodes, Douglas Thomas and, particularly, the late James Fisher.

I can only mention a few of the people who have made my travels memorable, especially in the West Indies and the Falklands, but these include: the late 'Dick' Barton, Jack Sollis and Ian Campbell, OBE, of the Falkland Islands; Mr and Mrs Carlos Felton, Killick Aike, Patagonia; Captain Bowden of HMS *Enterprise* and his crew; Lord Vestey for the loan of his cold room in Buenos Aires for my birds; Richard Dean of the Wildlife Society, Trinidad; Mr and Mrs Lau Bird of Paradise Inn, Tobago; Budram, our motor boat captain and guide through the Caroni Swamp.

An extra special word of thanks must also be given to Paul Nicholas for providing the exquisite line illustrations throughout this book, to Robin Woods for helping with information on the birds of the Falkland Islands and to my sister in law, Netta, for making those wonderful restraining jackets for the humming birds.

There are many, many others, too numerous to mention. Please accept my apologies if I have forgotten anyone.